W9-BST-017

TECHNO-CULTURAL EVOLUTION

TECHNO-CULTURAL EVOLUTION

CYCLES OF CREATION AND CONFLICT

William McDonald Wallace

Potomac Books, Inc.
Washington, DC

Library of Congress Cataloging-in-Publication Data

Wallace, William McDonald.
Techno-cultural evolution : cycles of creation and conflict / William McDonald Wallace.
p. cm.
Includes bibliographical references and index.
ISBN 1-57488-966-4 (hardcover : alk. paper)
1. Technology—Social aspects. I. Title.
T14.5.W36 2006
303.48′3′09—dc22
2005009078

Printed in Canada on acid-free paper that meets the American National Standards Institute Z39-48 Standard.

Potomac Books, Inc.
22841 Quicksilver Drive
Dulles, Virginia 20166

First Edition

10 9 8 7 6 5 4 3 2 1
26.95

CONTENTS

PART IV: OPENING THE AMERICAN FRONTIERS

PART V: TECHNOLOGY AND THE PROSPECTS FOR A META-CRASH

ACKNOWLEDGMENTS

Many people contributed to the ideas contained in this book, some of which date back to my Boeing days in 1960. But this specific book began to take shape as a result of an exchange of e-mail messages with Paul Ehrlich in 2000. He encouraged me take an evolutionary perspective on culture and economics. As my book began to take shape, Tim Flannery and Niles Eldredge gave me further encouragement based on a summary of ideas. Niles Eldredge was kind enough to read the whole of my first draft manuscript and gave me some valuable suggestions.

I also owe a debt of thanks for several semesters of students at St. Martin's University in my economic history classes for listening to my arguments, and then providing valuable feedback on an early draft. I incorporated many of their suggestions in my final draft. Three of my faculty colleagues were also of great help. Don Stout provided continuing counseling, feedback, and valuable suggestions from my first thoughts to the final draft. Norma Shelan also served as a "sounding board," and in addition she read the entire manuscript complete with annotations, often with thoughtful criticisms of some of my more controversial positions.

My thanks as well to Don McKeon of Potomac Books who agreed the day before Thanksgiving to publish this book. Thanks also to Michie Shaw for a marvelous job of editing.

But I must single out my wife, Patricia Ann Wallace, without whom I probably would not have finished the book. She gave me constant encouragement from the outset, was both a sounding board and devil's advocate, read draft after draft of my early writing (forcing me to discard much), and once the book was accepted, used her relentless "blue pencil" to edit the final manuscript. I owe her a world of thanks.

So, to all my deepest thanks, but any remaining failures are mine alone.

PREFACE

As a child growing up in the thirties, I looked forward with some excitement to the autumn unveiling of the next year's models of automobiles. Each new model seemed to be a big advance over the previous year and a four- or five-year-old car began to look antique. My bedroom walls were papered with pictures of airplanes for the rate of change in aviation was just as exciting to a ten year-old. A five-year-old fighter (such as the Boeing P-26) appeared to be a true antique next to the Lockheed P-38 or even the Curtis P-40. And indeed it was.

About this time, my teachers began discussing Darwin's theory of evolution and, unlike some of my classmates I had no trouble at all relating to it. (If a Spitfire could evolve from a Spad, I thought, why couldn't humans evolve from apes?) Although I regarded myself as a Christian, at age four I became skeptical about the literal truth of the Adam and Eve story. (If Cain and Abel were the children of Adam and Eve, and then Cain kills Abel, how in the world could Cain go off to a city to live? That, at least, is what happened according to the biblical story my Sunday school teacher had read to us. So, I asked myself, where had all the people in the city come from? Neither the Sunday school teacher nor my parents gave me a satisfying answer.)

In short, by age ten, I believed in evolution. After all, I could see new technology evolving before my very eyes. I could also see that as technology evolved so did our behavior. To kids my age about the worst put-down we applied to an adult was that he or she was still rooted in "the horse and buggy era" (that had passed away scarcely forty years before). Radios, telephones, movies, milkshakes, indoor plumbing, electric lights, and other appliances were all new but also were part and parcel of the modern age that we had been born into. We were utterly contemptuous of what had come before. None of us wanted to learn how to ride a horse. Instead, we wanted to learn how to drive a car. When we watched ancient Civil War veterans march (stagger, really) in the annual Memorial Day Parade we laughed at the displays of their weapons and uniforms compared to those in our "modern" army. We even looked down on that symbol of the modern era, the steam locomotive, for it could not compare to the newer diesel electric "streamlined trains."

As the years passed and technology continued to evolve, my sense of excitement dwindled but did not vanish. I became an avid reader of history during World War II and noted how armies had changed with technology. After the war, I took nearly as many courses in history as in my economics and business major. I would have majored in history but it didn't seem to lead to a job. My interest, in part, was a fascination with how our technology had evolved from crude stone weapons and tools to those of the modern world. But why was it, I came to ask myself, that despite "the Glory that was Greece" or "the Grandeur that was Rome" (chapter titles in my freshman world history text) these cultures introduced very little new technology. For the first five thousand years of all civilization, in fact, little new was invented. The big exceptions were converting from brass to iron and from pictographs to the alphabet.

Even more puzzling was why, despite the collapse of the Roman Empire under the barbarian invasions, and considering the Dark Ages that followed, did a now backward and nearly illiterate Europe suddenly begin inventing the new technology that was destined to put them on top. That technology after all (artillery, the printing press, the compass, square rigged ships and stern post rudders, the telescope, and much else) allowed the Europeans first to explore and then dominate the entire globe. Thanks to largely illiterate craftsmen, Europe acquired the power to leap from the most backward to the world's most powerful civilization. Why could not Rome have done as well at the very peak of its power?

This book attempts to answer those questions. But other questions also arose, and if any one event led to my writing this particular book, it was the "Battle of Seattle" during the World Trade Organization conference of November 1999. For the first time, I was forced to confront the downside of technology in a new and different light. Besides, my son Scott (an observer, not an activist) was mildly gassed during that event. Several of my colleagues from St. Martin's were among the activists protesting the conference just sixty miles to the north.

I long knew that the evolution of technology (via capitalism) was a process of "creative destruction." But I had celebrated that aspect. (Thanks be to God that the automobile had wiped out the buggy business. Otherwise I would never have had a car and I loved to drive!) I began to study economics at the very peak of John Maynard (by then Lord) Keynes' prestige, but in truth I felt Joseph Schumpeter, who coined the term creative destruction, was much more relevant since he focused on innovators and entrepreneurs. Whereas I knew traditional cultures often resisted American technology, I believed deep down that if only they knew better, they would happily embrace our technology and the modern democratic way of life that went with it. After considering the Seattle riots, however, I suddenly felt that assumption was just plain wrong. On September 11, 2001, I knew for sure it was wrong and I began serious research on this book.

For years I have lectured to my economics students that evolving technology drove all economic growth. I gave a thought experiment as proof. I handed around a stone blade I had brought back from Jordan that was said to have dated from about 15,000 BCE. Now imagine, I asked my students, what would your life be like if that blade was humanity's very best technological practice? We would all be nomadic hunter-gatherers living hand to mouth and 999 out of a thousand of us would not be alive today because that nomadic life style supports only six million, not six billion people. No one has yet tried seriously to dispute that conclusion.

In an off-hand way, I long understood that technology, economics, and culture all had co-evolved more or less together. When I decided to review Darwin's theory, a colleague from biology suggested starting with the latest modification to Darwin, called by its authors, "punctuated equilibrium." Going to the source, I read Niles Eldredge's latest book. He, with Steven J. Gould, coined that term. I had been familiar with punctuated equilibrium for some time but had read little about it since I had long accepted Darwin's theory. Still, I decided to bring myself up to date. I discovered for the first time that evolution was not the continuous process Darwin had said but rather had progressed by fits and starts.

When I finished Eldredge's book, I had a sudden flashback to 1960. That was the year I started at Boeing, just out of graduate school. A fresh young doctoral candidate (ABD), I was hired as the first graduate economist in the transport division. I spent about four months learning about Boeing's airline customers. Then in November the director of sales and marketing handed me an analytical hot potato. After annual growth rates of about 15 percent soon after World War II, in 1960 airline traffic growth had stalled. By September 1960, airline traffic volume was well below Boeing's forecasts. If that pause signaled market maturity (and some influential airline economists argued that it did), Boeing had grossly oversold the market on its new jets, the 707 series and an about-to-be-launched 727. Was the market mature, the director asked? Could I otherwise explain the pause just as the jets were coming on stream? Rushing back to my University of Washington professors (all Keynesians) I received no answers. They simply threw up their hands. Keynes was about depression, not growth, they pointed out. Remembering from my days at UPS that Schumpeter had stressed innovation, I studied the growth of every new industry I could find. I flesh out this story with more detail in Chapter 18 so here suffice it to say that I found a logical explanation for the growth pause. Nearly every new industry for which I could find data had pauses in growth between new generations of evolving technology. Therefore, I forecasted that growth in airline traffic would resume, and shortly it did.

But what I had more broadly discovered is that technology advances (or evolves) in fits and starts just as Eldredge and Gould said life did. Independently,Thomas Kuhn had found that science also advanced the same way going from a period of "normal" science to sudden paradigm shifts to a "new" science. I had long admired Kuhn's work but had not related his work to my Boeing study until I read Niles Eldredge. While reading Eldredge, I also read Tim Flannery's book on North America as the "eternal" frontier. Flannery made it clear to me how disruptions to stable ecosystems can trigger the punctuation phase of evolution and how behavior systematically differs in some important ways between life on a frontier of change versus life in a period of stability. That fact opened my eyes to the probable source of the intense animosity the United States was experiencing in much of the world and especially in the Middle East. The West's modern technology was forcing unwelcome changes that threatened sacred values and traditions. No one likes that, and indeed most of us will feel impelled to strike back.

With this background information, I felt I was on solid ground to pursue this project. It aims to fit biological evolution driven by random mutations in our DNA and cultural evolution driven by an analog, technological innovation, into a common Darwinian framework.

INTRODUCTION

Broadly defined, advancing human technology accounts for all of our economic growth and material progress. Yet evolution's path is littered with ruins. Technology must destroy even as it creates. Biological life itself has a similar pattern. Where there is life there must also be death, not only for individuals but for whole species. Well over ninety-nine percent of all species that have ever evolved have become extinct according to ecologists.[1] As Darwin and the evidence both make clear, the fit survive, the unfit perish.

Let's look at this process. Suppose someone has an idea about how to invent a new tool or weapon to solve a problem. Most such ideas don't work, but a few do, and if one works well, we embrace it. That means we must not only accept it, we must also learn new skills to use it. We must also specify the culture's "social rules of engagement" in regard to the new technology: who can use it, who can't, and for what purposes? What uses are off limits? These and other questions are often answered with rules or customs that involve new kinds of social relationships and sometimes new institutions as well. If we create new "hardware," we must also invent new social "software" to go with it. We embrace new technology by creating such software, but in the very same process, we usually undercut older technology and its software.

Technology's software makes up an important part of what we call "culture." Joseph Schumpeter, the early twentieth century Austrian economist, used the term "creative destruction" to describe the evolution of modern capitalism driven by innovators and entrepreneurs. Examples abound, and most of us have lived through and are living through this process. As autos displaced horses, as airplanes displaced trains for long distance travel, and as electrical appliances replaced domestic servants, the associated software changes of these and other innovations revolutionized our culture over this past century more than ever before. We gained much but much was also lost.

We are now working through this process of creative destruction with a series of new and more powerful electronic computers. Computers, in fact, gave us the term software. Bill Gates became the richest man in the world at a young age by realizing and then exploiting the fact that advances in computer hardware were not enough. The hardware needs software to provide it with an operating system. Without an operating system, a computer is simply a "tin box" as Bill Gates is reported to have said.[2] But once Gates's newly formed firm, Microsoft, had created MS-DOS and later Windows, the creation of software had just begun. Word processing, spreadsheets, data banks, systems of money management and accounting, along with many other applications, all emerged. Such applications induced both large and small adjustments in how we live with and relate to the new technology. The culture also had to adjust socially to reap much benefit from those new applications. All these changes induced new behaviors. For example, my own students now massively ignore the brick-and-mortar library in favor of online search engines and the Internet to research class projects. They can access research materials at 2 a.m. in their dormitories if they choose. They also spend great quantities of time on their cell phones when they aren't surfing websites, sending e-mail, or playing computer games. Such behaviors are a result of a huge cultural shift from my college days, and most of that shift has come since I retired from Boeing in 1992 and began teaching.

The broader rules of engagement include such things as social mores, customs, traditions, and language. The rule of law appeared early in the evolution of civilization. So did new economic systems and the methods of government. For example, hunter–gatherer small group egalitarianism gave way to the despotism of high output early agriculture and was followed again by the rise of democracy after the Industrial Revolution. Again, the software that goes with our hardware is what we normally call culture and even the broader term, civilization. All such software changes as we change our hardware. In the ancient world humans designed software (such as slavery) that retarded innovation to prevent change, as we shall see.

Darwin's Continuous Evolution

That said, we need to look more closely at the process of evolution as first outlined by Charles Darwin in 1859.[3] Mostly as the result of a long voyage of discovery on HMS *Beagle* in the 1830s, Darwin concluded that life changed as time passed. He called such changes "evolution." It had long been obvious that as a general rule "like begets like." Yet every now and then new characteristics would appear randomly in both plants and animals, and these changes were in turn passed on. This observation led to the selective breeding of both plants and animals, which produced major modifications to the plant and animal food we now eat compared to their original state in the wild. Selective breeding has been undertaken by humans for thousands of years. Darwin hypothesized that random mutations selected for breeding was a variation on a process of "natural selection" that determined whether any given mutation would "take" and be passed to the next generation. The mutations would take, Darwin concluded, if they provided additional fitness for the individual in life's game of competition for space and scarce resources in its natural environment. If the mutation did improve fitness, then it was likely to be passed on. The main criterion of fitness was the ability to survive to the age of reproduction and then pass on one's genes (although Darwin had no modern knowledge of genes.) The whole process was summed up as the "survival of the fittest through the natural selection of random mutations." The great majority of such mutations would not take, Darwin acknowledged, but in the fullness of time such a process created the new and more diverse species of plants and animals that he called evolution.

Enter Punctuated Equilibrium

Darwin assumed evolution was a continuous process in good part because he began his research believing in the principle of "uniformitarianism." This principle dominated geology from Darwin's time to the mid-twentieth century when it was displaced by today's theories of shifting plate tectonics.[4] The scant fossil evidence, however, did not support Darwin's hypothesis of continuous process. He knew this, but Darwin argued the fossils were too scattered and few in number to prove anything. He relied on the logic that random mutations were always taking place and thus natural selection would assure a continuous process. Besides, Darwin and his supporters insisted that as fossil evidence accumulated, it would sustain his hypothesis.

In fact, it did the exact opposite: Over the next one hundred fifty years a large accumulation of fossil evidence cast increasing doubt on Darwin's continuous process of change. In the 1970s, two paleontologists, Niles Eldredge and Steven J. Gould, showed convincing evidence that evolution moved in an erratic fashion of bursts of actual change often followed by much longer periods of stable equilibrium or stasis. Eldredge and Gould named this step process "punctuated equilibrium."[5] As a "step function," punctuated equilibrium is similar to other natural processes such as plate tectonics. Here, pressure between separate plates builds up over some period and is suddenly released in the form of an earthquake as the earth's plates suddenly release and slide past each other. The earthquake is the punctuation; the period at rest is equilibrium.

Fitness, meanwhile, determines both individual and cultural survival. The culture's combination of hardware and software make up the "techno-structure" that determines how well a culture can compete in its environment. That environment will include the other cultures competing for the same resources in the same environment. As techno-structures evolve to become more complex, differences in the relative "fitness" of techno-structures and hence cultures grow wider. However, fitness is a relative term. Something is fit or unfit relative to the environment in which it must cope. A sudden shift in the environment (and new technology can itself cause a shift) can change the relative competitive fitness either of a person or culture. A culture's internal techno-structure serves a purpose similar to that of the DNA in any living being, including humans. Both are repositories of relative fitness. Moreover, the evolution of that techno-structure follows rules similar to those that govern the evolution of our DNA, except that culture can evolve far faster than can DNA.

Here we need to understand the broad pattern of evolution itself because that pattern and its rules apply to both human anatomy and culture. Once we understand that evolution changes both our anatomy and our culture by the same process, we open the door to a unification of the social sciences with the biological sciences. That is similar to what Darwin's original theory was able to do. Darwin unified botany and zoology into a common science of biology. Commonality means that the DNA both of plants and of animals follow similar rules that govern how changes in their respective DNA evolve. Darwin's original theory could not address the evolution of culture because he assumed evolution was a continuous process. Since his assumption was incorrect, we need to take a closer look at why and how the changes take place in the "punctuation" phase of punctuated equilibrium. We do that in Chapter 1. First, however, let us briefly review the kinds of problems that the rapid pace of today's cultural evolution of our techno-structure is now creating for us.

Circuit Overload and Serious Dilemmas

Many of our cultural divides arise from the stresses caused by "creative destruction," by which process technology drives cultural evolution. We now invent new technology so fast that it creates serious "circuit overload" in human cultures, and these trigger a host of nasty problems the world around. These problems come entangled in some perplexing dilemmas. For example, we often invent new technology to solve existing problems and it often does so very well. Six thousand years ago, the invention of wheeled carts hauled by domesticated animals solved a big transportation problem. Oxen pulling carts allowed supplies to cities with food. Yet those same oxen or other draft animals soon created urban pollution—filthy streets that stank from the urine and feces of draft animals. Later, automobiles solved that and other problems so well that hundreds of millions were sold. But the exhaust gases from those millions of cars and trucks created a huge problem of urban air pollution. Currently, tens of thousands of people die from auto accidents year after year.

In short, we invent technology, and then embrace it. We begin to adjust our way of life to the new technology by creating the social software that allows us to engage effectively with the new hardware. Suddenly, we find ourselves dependent; we can't do without this new techno-structure. It is then that we often discover serious delayed side effects induced by the large-scale use of successful innovations. We can't turn back, so we invent more technology, and the social software to go with it, to solve those problems. The cycle repeats itself. In American culture, the pace of change seems to pick up with each cycle. So does circuit overload.

Two important issues stand out. First, the circuit overload induces a backlash in traditional ("Third World") cultures that are in equilibrium against high-tech ("First World") cultures still in the punctuation stage. The traditional cultures fear, quite correctly, that technology of industrialized nations and its cultural software, pose serious threats to their traditions, values, and religion.These things often define the core identity of those living in a tradition-focused culture. In short, new technology directly and indirectly threatens their whole way of life. Thus, a major clash between the two types of cultures has erupted. High-tech has imposed itself on low-tech cultures for at least eight thousand years. But until recently, the pace of change was much slower, far less intense, and took place mostly on a local rather than global scale. The end of the Cold War, however, did two things to speed up and intensify the process on truly a global scale. First, a surge of new electronic micro-chip technology emerged including powerful personal computers, more sophisticated software, and the World Wide Web to go with them, global positioning systems, cell phones, and much else. Second, the collapse of the Union of Soviet Socialist Republics (USSR) dissolved a major buffer. The contest between the USSR and United States had allowed traditional Third World cultures to play off the two adversaries, providing them with some control over how fast and in what way they would adapt new technology. When the Cold War ended, it seemed clear that the software of capitalism favored innovation and entrepreneurship far better than did communism, a conclusion the Soviets themselves drew. That "victory" seemed to unleash a sudden unconstrained American passion to globalize capitalism, free trade, universal individual rights, and liberal constitutional democracy. These institutions all comprise important parts of the software that sustain the self-accelerating technology in Western cultures. But the pace of that same technology often does much violence to the values that tradition-focused cultures hold dear.[6]

While technology continues to define western civilization, traditionalists fear the West's soulless and impersonal market competition forces ever newer high-tech devices and software upon them: "Download or die," as Thomas Friedman put it. This package is perceived by the Third World as a toxic set of temptations that threatens to dissolve their sacred traditions and strikes at the heart of their higher core identities.[7] Such fears did much to motivate a terrorist counter-attack by Islam, the civilization that sees its values most threatened by America's modern techno-structure.[8] The attack strategy of zealots who espouse Islamic terrorism partly outflanks our technological advantage. Terrorists can acquire much of our technology to use against us and under cover of our democratic rights and freedoms. They can even use our legal code of due process to protect their privacy as they plot against us on our own soil. Secure privacy was, in fact, a major factor in their success in destroying the twin towers of the World Trade Center on September 11, 2001.

When America struck back, as in Afghanistan and Iraq, we discovered our well-trained, high-tech military could disable their conventional forces by "shock and awe" and with low American casualties. But once the region was in occupation, a dreadful dilemma emerged. Much of the population remained rooted in local traditions and continued to fear us. Guerrilla warfare then is very difficult to subdue in the context of Western democratic values. Tamerlane, who conquered this region in the fourteenth century, had no such trouble.[9] At the first sign of resistance, he "put the city to the sword" and thus secured his position by killing everyone. Western values preclude such tactics.

Unsustainable Growth

A second major issue entered our awareness only about thirty years ago, namely the threat to the global ecosystem created directly from the great success of our technology and its software, including capitalism. It created a vast increase in wealth. That wealth has been widely distributed within the high-tech cultures that created it, despite frequent claims to the contrary. Had this wealth not been widely distributed, there could be no mass consumption binge. This binge–and the pollution that comes with it–put the integrity of our global ecosystem at serious risk. But our awareness of the threat to the ecosystem began less than one hundred years ago. Up to about 1900, about 80 percent of the population of nearly any country lived in what we now define as grinding poverty. Poverty, in fact, was the lot of the great majority of people from the dawn of civilization until 1900. Then, in about one hundred years, real per capita income shot up more than ten-fold in the high-tech nations as a direct outgrowth of America's rapidly advancing techno-structure.

Of the one hundred thousand-fold total increase in consumption since humans began to farm about ten thousand years ago, at least half of that total increase came after the end of World War II. The world's population has jumped from six million to six billion, but two-thirds of that population growth came in just the last fifty years. The huge amount of food we must now produce to feed everyone draws out ground water at an alarming rate in America's own Midwest "breadbasket." Moreover, the technology behind the Green Revolution that so increased food output in Third World nations requires a monoculture, or single species, that is far more vulnerable to destruction by diseases resistant to herbicides. Mixed species crops do not face that risk but are often much less productive.

A Tough Choice

Still, can't well meaning people from various nations get together in the common interests of humanity and agree in good faith to put a stop to runaway technology, consumption, and its pollution? Perhaps, but let's first look at a few of those perplexing dilemmas that I mentioned earlier. For example, about a third of the workforce in the developed nations–and much of it in other nations as well–now depends for their affluent livelihood on a sustained growth in the consumption that creates the pollution. After all, my consumption binge may provide you with your livelihood. Constrain growth and the specter of mass unemployment arises. That specter for Americans first arose in the Great Depression triggered by the temporary maturation of several technologies such as the automobile before such newer technologies as aerospace and electronics grew large enough to restart the growth.[10]

But there is a more serious political dilemma. The current increase in the pace of new technology began with the rise of market-based liberal constitutional democracies. In such democracies the people are sovereign and not despotic. Constitutional democracies emphasize individual rights and personal freedom, and the United States broke new ground on this idea with its Bill of Rights.

These rights apply to innovators as well as to entrepreneurs and they, not old established firms, create growth industries based on new technologies that create new and more affluent jobs. These rights and freedoms are vital to innovation. Without them, the more mature industries would soon quash innovation to maintain their market position. That the mature firm was once an innovative start-up becomes irrelevalent. America led technological advances in the past, and now the force of government prevents large, mature firms from constraining newer, but much weaker innovators and entrepreneurs. The average citizen expects to have the right to innovate, even at the possible compromise of the global environment. Even environmental activists would be outraged; they depend on those very same rights to mount such protests as the World Trade Organization riots in Seattle in 1999.[11] Environmental activists are caught in a true dilemma.

The post-Cold War clash of cultures demonstrates this dilemma. As mentioned earlier, our freedoms and openness enabled our adversaries to plan and launch the attack of September 11, 2001, under full cover of American protections of due process. Still the government was soon criticized for not having foreseen and stopped the attack. It is, after all, charged with the national defense. Consequently, Congress passed the Patriot Act that curtailed some due process protections and created a Department of Homeland Security for oversight. Enormous outcries began at once. Americans do not want their freedoms abridged. They fear government agencies communicating with each other about predetermined suspicious behavior or privacy matters such as individual choices. With good reason, people fear that such power sooner or later will be abused.

In sum, to abridge those freedoms to the point of quashing new technology would also quash the American Dream of affluence as well as our democracy that we have come to know and love. We would also incur mass unemployment, because if new technology slows to a stop, so does the growth of jobs and consumption. Indeed, the number of jobs drops from growth plus replacement demand to replacement demand only. Abridgement of rights also quashes the right of protestors to demonstrate against whatever they want to resist. For quite different reasons both the political right and the political left would resist any serious erosion of those rights.

If we want to deal intelligently with these issues, we need to understand why and how technology drives cultural evolution. So, let us turn next to a more detailed look at how the process of evolution takes place and how it induces different kinds of behaviors. We will begin with a detailed discussion of the punctuation phase of evolution. We will see how disruptions to stable ecosystems trigger punctuation by creating a new frontier of change that, in a five step process, works its way back to equilibrium, but not until considerable evolution takes place. I call this five-step process FROCA—it begins with a frontier (F) and ends in an adjustment (A) that brings forth a new equilibrium.

The rest of the book carries the FROCA process forward. We begin at the time humans came down from the trees, move on to life as hunter–gatherers, then to agriculture and despotic ancient civilizations that ended with the collapse of Rome. We then track the resurgence innovative technology made possible by the fall of Rome, to the European Enlightenment and rise of democracy, followed by America's emergence as the sole global superpower. In the final two chapters, we address some possible global futures.

This book also includes two appendices. The first looks at the new science of quantum physics (that came after Darwinian Theory) in relation to teleology. Teleology is a philosophical term that refers to a higher purpose. Does evolution follow a divine or a higher cosmic purpose? Or has evolution followed a purely random path as Darwin and most scientists since have believed? This has become a hot debate as newer sciences cast current light on the question, reopening the debate with science. I merely point out that teleology is now an open scientific question. The reader is left to reach his or her own conclusion concerning evolution's higher purpose, if any.

The second appendix considers three modifications to biology (including punctuated equilibrium) that emerged in the 1970s and the new science of Chaos Theory that emerged at about the same time. Jointly, the four hold promise of crafting an integrated general theory of human behavior that unites the social sciences with biology. Here the focus is on the new concept of "autopoiesis," a coined term from Greek that means "self-making."[12] In essence it is an extension of the "law of self preservation." Along with Chaos (or Complexity) Theory, autopoiesis allows us better to understand the long-noted and widespread phenomenon of self-organization. It seems to refute mechanistic randomness but does not demand a teleological explanation. Again, autopoiesis governs certain kinds of behavior at all levels of life. Thus, its application to humans and human culture opens new doors to our understanding of human behavior, both by individuals and by groups.

PART I

FROM RISEN APE TO HUMAN TECHNO-CULTURE

1

THE FROCA PROCESS OF CULTURAL EVOLUTION

Human anatomy and its techno-culture coevolved for about two million years and perhaps longer.[1] The rate of change was slow at first, with long periods of equilibrium between punctuations. Gradually, human inventions began to so amplify the power of our anatomy that for the past thirty thousand years, our new technology has all but preempted the ability of genetic mutations in our DNA to convey additional fitness. Our inventions in technology open up new frontiers that allow accomplishments and expansions we otherwise simply could not imagine.

The pioneers moving forward on such frontiers feel a sense of "ecological release" from the previous constraints under which they lived. This release allows them to change their behavior. And change they must, not only to cope with new risks but also to take advantage of the new opportunities to be found along the new frontier. They are, after all, pioneers exploring unknown territory. In the process, new circumstances redefine what fitness implies and for humans, may even modify ethical standards. Previous standards often seem irrelevant, sometimes counterproductive. They may inhibit the pioneers from exploiting new opportunities or from coping with new frontier risks. Pioneers must move fast to take advantage of the frontier's new opportunities if they are competing with others. Be quick or lose out.

In this competitive struggle, good results trump an emphasis on following proper procedures common in more stable pre-frontier communities where cooperation rules. As a group, the pioneers overexploit the opportunities that ecological release created. If the opportunities emerge in the form of new technology, the pioneers discover, sometimes with a sickening suddenness, that the new technology has matured. Maturity can extinguish the opportunities that growth provides, sometimes in a flash. This abrupt downshift puts some pioneers out on a limb of overexploitation, a bough that often breaks. After an opportunity's growth crashes, some heroes of the frontier may go from hero to villain almost overnight, as if they are the purveyors of the maturation, not creators of the technological advance. In any event, the least fit or unlucky pioneers don't survive the crash. The survivors often must make a painful readjustment to the new post-crash environment, much as did the surviving dot.coms after 2001.

The new technology that opened the frontier will remain in place, of course, and indeed the culture may have become dependent on it and its software. Some of the new behavior it permitted will also survive the crash. But the now-mature technology is no longer typified by rapid improvements. The go-go days are over and the now-mature technology pressures survivors to follow more conservative standards of behavior. Thus, a post-frontier set of standards, rules, and values emerges to signify that the culture has reached a new equilibrium. The culture may revise some software to slow the pace of change. Stasis returns. The culture settles down and remains stable until newer technology or other major disruption upsets the ecosystem's equilibrium.

Each new disruption creates chaos from which a new frontier emerges. Disruptions, remember, create the new frontier. Only then comes the ecological release from any earlier constraints. Release both attracts new pioneers and energizes those simply present during the disruption. Again, pioneers are eager to take advantage of release to exploit the new opportunities as fast as they can. A competitive rush begins that later ends in overexploitation, followed by a crash. The survivors must again adapt to yet another post-frontier, more cooperative equilibrium. In sum, after each major disruption the five-step sequence begins: Frontier, Release, Overexploited opportunity ending in a traumatic Crash followed by a sometimes painful period of Adaptation (FROCA).[2]

The FROCA process can happen on both the macro and micro levels, in particular industries, in subcultures, or for whole cultures. But this five-step process is not unique to culture; it also applies to biology and unpacks the punctuated part of "punctuated equilibrium." Punctuation incorporates the first four steps, F.R.O.and C. Equilibrium comes as a result of a final step, the A of adaptation.

The mutations that drive biological evolution are more possible on a frontier rather than in an equilibrium. They spread much more slowly than new technology because mutations are constrained by the limits of sexual union. Mutations happen to both men and women. But before mutations can spread, the mutated individual that got a mutation must mate with a person of the opposite sex. Each partner contributes only half of the DNA passed on to their offspring. Thus, the mutation itself may or may not be passed on. Genetic dilution puts a real brake on the propagation of any mutation from one generation to the next. Mutations spread fastest in small, isolated populations since small size reduces the constraint of genetic dilution. Disruptions often isolate members of the gene pool, such as with the tectonic plate shift that split South America and Africa apart. Mutations happen all the time, but even beneficial mutations are often diluted out of existence in a large gene pool. The same mutation, however, may take hold and propagate in a smaller gene pool.

Disruptions often change the ecosystem's criteria for fitness. Thus, a mutation in a small, isolated group improves genetic fitness in the new and changed environment and is much more likely to take hold and propagate. That is why new species usually emerge during the chaotic, but also creative frontier period following a major disruption.

Unconstrained by sexual union, technology, however, is

not limited to a propagation process that can go no faster than one generation at a time and can be diluted out of existence in the process. New technology often spreads faster in larger populations. If anything, a small, isolated population tends to limit the propagation of innovations. Historically, mutations, complete with their propagation constraints, drove nearly all evolution before they evolved the kind of DNA that led humans to invent the technology that in turn drove their own cultural evolution. Thereafter, technology would itself become a major source of the disruptions to the ecosystem that can trigger the FROCA process into an ever faster positive feedback loop of self-acceleration. Some scientists argue self-acceleration puts human evolution on a dangerous path where overexploitation courts terminal risks.

Whether or not that view holds up, several million years passed before human inventions in new technology actually displaced mutations. Our distant ancestors came down from the trees about six million years ago. They still had small brains, but their small group mutations had given them an upright stance and opposable thumbs. Slowly, they learned how to "craft" crude stone weapons. That primitive technology favored the evolution of larger brains once the disruptive Ice Ages began. Larger brains led to the invention of more sophisticated Stone Age technology and a growing ability to adapt to colder climes after humans learned how to make fire. Before long, our new technology began reinventing us, slowly at first but then with gathering speed. The early changes were mainly in our anatomy, but these enabled us to invent more complex technology. Ultimately the technology enlarged the cultural possibilities allowing for a greater variety of complex cultures both in geography and in terms of social, economic, political, artistic, and religious expression. For the last thirty thousand to sixty thousand years practically all human evolution has been cultural, via technology, and not genetic, via mutations. For example, stronger arms made little sense after we invented the bow and arrow that amplified the strength of our existing muscles.

Once we domesticated horses, we could use their far stronger legs to travel three times as fast as our legs could carry us. Thus, a mutation for stronger legs might not convey more relevant fitness. What good would incipient wings be when we can invent them ourselves to fly in 747s more than ten times as fast?

The relative fitness of individuals, and indeed groups, in almost any endeavor now depends heavily on what technology they command and how well they can use it. The fastest runner can't compete with an indifferent runner driving a car. Strong cultures invent strong technology, become skilled in using it, adjust their modes of living to take advantage of it, and incorporate the legitimacy of such use into their system of values. Historically, the cultures that have done this the best have, in the event of conflict, either taken over or driven out cultures with less fit techno-structures. Professor Jared Diamond points out that farming cultures always displace hunter–gatherers from good farmland.[3] Farmers either drive the hunters away or absorb them into their own culture because the techno-structure of agriculture allows the land to feed more people by far compared to hunting and gathering. One exception was when the Inuit pushed the Vikings out of Greenland about 1500 CE. But that was during the little Ice Age when a much colder climate reduced the efficiency of farming in Greenland. The Vikings migrated to that big island about 800 CE during a comparatively warm spell.

Few exceptions aside, the story of civilization from the outset is a story of strong techno-cultures displacing the weaker ones. More than technology is at work, to be sure, but without the new technology—beginning with agriculture—civilization could never have arisen. Again, by thirty thousand years ago, the action in human evolution had shifted almost entirely to culture. Paleontologists have chosen that period because between sixty thousand and thirty thousand years ago, human culture suddenly took its first large jump in technical sophistication after hundreds of thousands of years of gradual increase.[4] Moreover, major hominid variants had vanished by then.

During that long and rather gradual transition human anatomy and technology appear to have coevolved interactively, as mentioned earlier. It is clear that the modern sized human brain had evolved by perhaps half a million years ago. No one knows just what caused the sudden jump in new technology. One plausible theory put forward by paleontologist Niles Eldridge holds that when our anatomical "software" enabled us to speak in complex languages, techno-culture took over, because language promoted the more rapid spread of new technology. Still, we have no immutable proof that modern language evolved abruptly.[5] Perhaps the important point is not so much what caused the sudden shift as the fact that it took place. Ever since, new technology and its cultural software has dominated human evolution.

Not long after the great leap forward, humans became such effective hunters that they began to overkill the game. In the process, they wiped out the majority of the very large temperate zone mammals, such as the woolly mammoth, mastodon, and saber-toothed tiger, among others. A crash of sorts occurred when the game vanished in local areas. Thrown back to gathering, humans learned how to cultivate and domesticate various crops and began systematic and settled farming. No longer nomads, humans could live a settled life, store surplus food, domesticate animals to eat and others to take over much of the heavy lifting. In these circumstances, humans had the time to practice a much greater division of labor that encouraged a surge of new specialized technology, and thus ancient civilizations based on cities became a practical possibility.

Once fairly large cities (more than thirty thousand people) emerged, new technology all but shut down for the next five thousand years, as we shall later see.[6] Then in 476 CE, the Western Roman Empire collapsed. The collapse of Rome soon devastated Europe's social order and caused the Empire to fragment into tiny local principalities. By about 700 CE, the Dark Ages had arrived.[7] Rome's collapse was later to provide ecological release from its centralized political authority to craftsmen who discovered opportunities to innovate. In a fragmented world without central political control, craftsmen found a new and creative voice in their own small towns where most of them lived. Slowly at first but with gathering speed, craft innovations began to mount. By 1500 CE, a "punctuated" explosion of new technology emerged. Western Europe became a new frontier of better technology that enabled Europeans to spread around the globe. They soon dominated it. Along with its new software, new technology transformed human cultures at an evermore rapid pace of "creative destruction." By the end of the twentieth century, that frantic pace had induced serious psycho-circuit overload in many cultures.

Now, because no political authority has the power to stop it, some scientists fear that our rampant new technology with its high rates of consumption and massive collateral pollution will devastate the globe's self-adjusting ecosystem. To repeat the statistic giving rise to that fear: Human consumption increased one hundred thousand-fold over the past ten thousand years, and 60 percent of that increase occurred in just the past sixty-odd years. Again, a vast increase in pollution of all kinds is part of that growth and it cannot long continue without some kind of crash. Humans have caused many species to go extinct in the past. Paleontologists have identified five major extinctions, and some biologists claim that humans have already caused the sixth wave of major extinctions.

Recall that punctuated equilibrium divides evolution into frontier periods of change (punctuation) followed by more stable periods (equilibrium). This book draws on a related theory developed mainly by ecologists such as Tim Flannery who unpack the frontier effect into the five-step FROCA process.[8] These ecologists point out how major disruptions to stable ecosystems upset the balance and induce chaos to create the frontier effect. A major disruption may spell disaster for some of the inhabitants, even extinction. For example, the dinosaurs went extinct about sixty-five million years ago after an asteroid hit the earth near Yucatan by the Gulf of Mexico.

Still, past disruptions and the ensuing chaos often released its survivors from the very constraints that brought balance and stability to the ecosystem. For example, the Yucatan asteroid hit that wiped out all dinosaurs, released mammals from the serious constraints on their evolution imposed by dinosaur dominance. Those constraints suddenly gone, whole new opportunities for expansion and evolution opened up for mammals.[9] Mammals suddenly found an open door for expansion and branched into a more diverse family of species. Reflexively, life usually takes advantage of any opportunities to expand. Were it otherwise, evolution would probably not have begun.

From the expanding and evolving mammals, the apes appeared. Some of those apes evolved into upright hominids, and hominids evolved into modern humans. The specific path of human evolution was triggered by a series of major climatic disruptions to a previously stable tropical ecosystem in Africa that culminated in seventeen consecutive Ice Ages.

That disruptions give rise to new opportunities continues to be a part of current everyday life. However devastating, a disaster often provides ecological release from previous constraints. Such release can open up new opportunities. For example, the policy failure of the reigning political party creates an opportunity for the opposition party. If a much-beloved leader in business, politics, or the military dies, that loss opens up opportunities for others, who were previously constrained by that leader's success.

On a messier note, floods, fires, and earthquakes can introduce temporary ecological release from law and order thus creating "opportunities" for people to loot. Indeed, we can see temporary ecological release from prior constraints happen all the time. In the aftermath of victory, soldiers historically went on rampages of plunder and pillage. Sailors, after long months of confinement and constraint under sail at sea, went on drunken binges in port. Staid business executives at conventions who discover release from their daily grind sometimes "go ape." Both men and women have taken lovers who offer occasional release from the constraints of a difficult marriage. Young people, who leave homes that featured tight discipline and strict rules, may also "go wild" once they discover freedom in a new setting. A stock market boom like that of the 1990s dot.coms created the prospect of ecological release from financial constraints that tempted many people to take big risks. Some ended up in bankruptcy.

The dot.com bust of the 1990s, the gold and silver bubbles of the early 1980s, and the stock market crash of 1929 all illustrate the point. Such crashes or busts almost always follow big booms. They often cause great trauma, not excluding extinction. A major kill-off, figuratively or literally, follows. The survivors struggle to adapt to a new set of emergent constraints, including constraints on personal freedom. Still, these crashes were and are a vital part of the punctuation phase of evolution. Only after pioneers exhaust the opportunities of their new frontiers can the ecosystem establish a new balance and then adapt to a new equilibrium. By the time it does adapt, that ecosystem itself has evolved. A major shift in the code of ethics occurs from the relative looseness of the frontier to a much stricter code during stable equilibrium. Moreover, the frontier's pragmatic concern for results by nearly any process that works ends. In a stable cultural ecosystem, the emphasis shifts to a concern for a "proper" process that must be followed regardless of results. The frontier focus on a growing pie shifts, in equilibrium, to a more fair distribution of a constant pie. These two goals require quite different policies, and that fact generates much debate.[10]

This process of Frontier, Release, Over-exploited Opportunity, Crash, and finally Adaptation is the way that FROCA links punctuation to equilibrium following the disruption.

While today's technology is self-accelerating, it was not so in any ancient civilization. At most there was a slow upward drift. Nearly all ancient civilizations were despotisms that in various ways sought social stability. Stability was usually achieved via strict social obligations, precise rules of conduct, and deference to one's social "betters." Those who resisted were punished, even executed. Innovations were not welcome. They upset this social balance and so were discouraged, actively suppressed, sometimes abolished. After the fall of Rome, much of the culture's technology fell into disuse and was temporarily lost.

Worldwide, the shift to a democratic political system based on personal and individual rights created new political "software" for an emergent industrial enterprise system that gave rise to self-accelerating technology. For instance, in 1776, the United States became independent—a new nation with a government that for the first time in history made the people sovereign and enforced a Bill of (personal) Rights to make it so. That act let the technological genie out of its bottle long corked by ancient despotisms, and that genie is still at large. In short, America's Bill of Rights was the most important piece of software ever created to promote a continuous process of innovation and new technology. Look at the results: A fledgling new and sparsely-populated coastal nation morphed into a continental power, and 125 years later in 1900 was also the world's largest industrial power. Ninety years after that, the United States became the world's only superpower made possible mainly by its rapidly evolving techno-structure.[11] The software we call the Bill of Rights made America's record of growth in economic, military, and political power possible.

However, such power made the United States a prime target for widespread resentment. Envy was only part of that reaction. Fear also played a major part because the more tradition-centered cultures feared America's open "frontier ethic." They feared its culture (software) of constant technological advance that often imposed unwanted changes upon them. Many and perhaps most people prefer to live a stable life with its constant values and stable social relations held in place by traditions and resist a major change unless it offers an immediate reward. Thus, many people in other cultures see the United States as a rambunctious high-tech culture that poses a serious threat to the stability of their values, religion, and often, their sacred traditions. America is seen as a sort of barbarian cowboy super-state trashing social ecosystems all over the world. A number of Americans feel the same way.[12] Many academics promote this view, illustrated by their much different reaction to September 11, 2001, compared to the majority of Americans. The cultural rift has opened up in America as well.

Still, new technology creates new opportunities even in tradition-based cultures that often appeal to young people, as author and columnist Thomas Friedman points out.[13] New technology can offer ecological release from the constraints of tradition, especially to younger people. They see a chance to escape the tyranny of their elders, dismissed as "old fossils" who can't grasp the new technology that young people quickly learn and then want to use. Indeed, the early adapters of new technology in almost any culture are usually younger people. That fact, of course, greatly alarms the older folks, as Friedman also makes clear. Thus, the latest new technology will always offer a competitive advantage to those nations convinced they need an advantage to preserve their core identity as a nation. Otherwise, it may take some dreadful crash that invites a return of despotism to constrain technology and restabilize the global social and physical ecosystem. That first happened with the rise of civilization in response to anarchy induced by high-output agriculture's population explosion. The prospect of a similar event happening in the future as a result of the recent consumption explosion is explored in the book's final two chapters.

All life has twin inbuilt tendencies that pull in opposite directions, to expand and yet to remain stable. That duality is the essence of punctuated equilibrium. Both tendencies are driven by an impulse called autopoiesis. As mentioned earlier, autopoiesis is an aspect of the law of self-preservation focused on identity. But the requirements for such identity-preservation work differently on the frontier of change compared to a stable community. On the frontier, people tend to exploit any opportunities they can find to get as much as they can and grow as fast as they can. For example, the eighteenth-century philosopher, Thomas Malthus, built his theory of population growth on our urge to reproduce. Further, most people want to live in a stable and reasonably harmonious ecosystem. The first four steps of the FROCA process deal with our drive for growth in the punctuation phase. Adaptation, the last of the five steps, brings out our human desire for a stable life in equilibrium, after we adjust to it. The alternation of these two drives of growth and stability completes the life-cycle of evolution defined as punctuated equilibrium.

2

CLIMATIC CRISIS AND THE RISEN APE

About seven million years ago, the climate in Africa changed. It became somewhat cooler and much dryer. This dryness disrupted the ecosystem of the tropical rain forest at its margins. There, rain forests began to shrink, and as they did, a new ecosystem evolved to adjust to the now dryer climate. The new ecosystem was a far more open savannah, a mixture of more grassland and fewer trees. It soon became habitat for many species of grazers and browsers that rarely ventured into tropical rain forests. We don't have a complete record of events that took place during this shift and readjustment. Still, we do know that a new species of primate evolved in that process. This new species became our earliest ancestor, the one that set us apart from the chimpanzee, our closest primate cousin. We share 98 percent of our DNA with the chimps, according to Jared Diamond. We call this ancestor a hominid. Modern paleontologists have unearthed many fossils of the several hominid branches over much of Africa.[1]

What set this new creature apart from the chimpanzee was its ability to stand upright, as we do. It also had hands with opposable thumbs, much like ours today. It was, however, no smarter than a chimp, judging from the size of its brain case of about 350cc, the same size as today's chimp. Yet this risen ape, the hominid, began to develop a new lifestyle as an adjustment to its new ecosystem, a lifestyle that would not have worked well in its previous ecosystem of a dense tropical rain forest. The shift to this new way of living apparently took place over a period of hundreds of thousands of years. But by about six million years ago (mya) this hominid had evolved into a creature that was to remain relatively unchanged until about the time the Ice Ages began just under two mya. Such stasis is an example of punctuated equilibrium, a relatively rapid evolution of anatomy followed by a much longer period of stability or equilibrium when little change takes place.

What induced some tree dwelling chimps, probably on the margins of the rain forest, to evolve into upright hominids? Why didn't those changes take place among chimps in general? Here we must depend on informed speculation. Most paleontologists today believe that as the tropical forest became dryer, predators such as hyenas and lions made incursions into the emerging savannah. Although a chimp can live perfectly well in a savannah, survival depended on moving fast enough to escape the new predators. Those that couldn't escape perished. But for a few of those chimps, mutations in their DNA allowed them to stand more upright. Uprightness allowed them to see predators coming sooner and thus escape. Not only could they see much farther, they could also run, and thus move faster. Such mutants survived. They passed on their modified genes, or anatomical innovations, to their offspring. Their offspring survived to reproduce more often than did those without this innovation. Thus this mutation was able to spread quickly where there was a relatively small population of local animals. Beneficial mutations always spread fastest within small relatively isolated groups cut off from the main body of their kind by the disruption.

That isolation also explains why the chimps that continued to live in the now smaller tropical rain forest did not become upright. That mutation would not have provided more fitness in that ecosystem. Conversely, it could have reduced fitness, because creatures with an upright stance couldn't escape their predators by swinging through the trees with the facility of a chimp. In short, fitness is not abstract. A change in anatomy (or as we shall see later, in technology) adds fitness or not relative to the animal's habitat and ecosystem.

Another advantage of the upright innovation in the dryer

ecosystem came in connection with the opposable thumb. Together, these changes later allowed the hominids to craft crude stone (and possibly stick) weapons, which they probably used to defend themselves against predators. Still, by about two mya, as the climate again became suddenly dryer and savannah more open, the hominids had created a significant inventory of crafted stone tools or weapons used to hunt game. These are called Oldowon tools as illustrated in Figure 1.

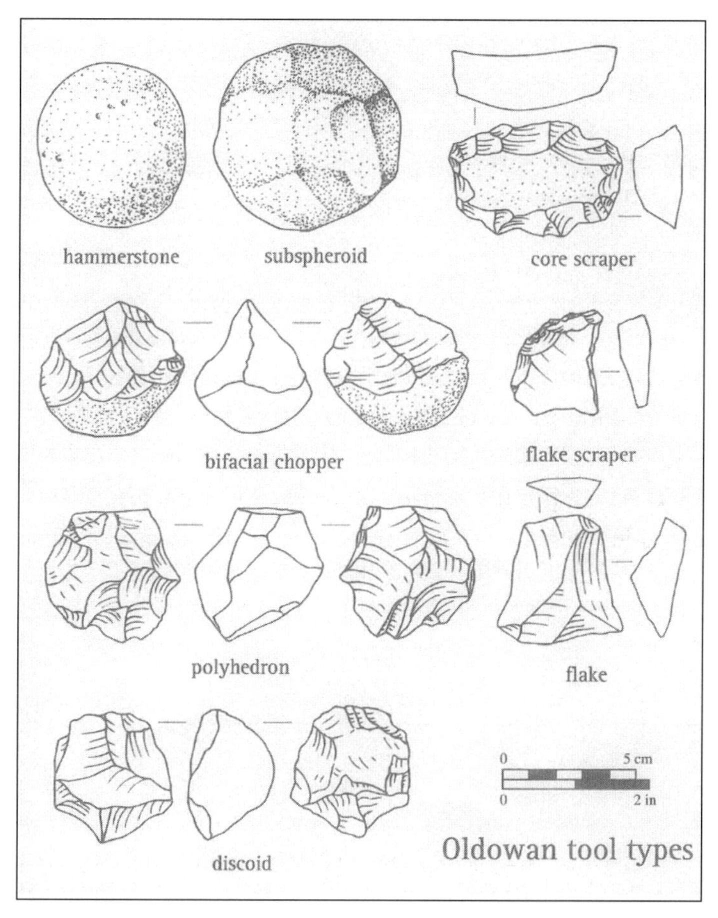

Figure 1: Representative types of Oldowan stone tools, *recognized by Mary D. Leakey and other specialists, from* The Dawn of Human Culture, *Richard G. Klein (reprinted with permission).*

In effect, the newly evolved hominid joined other predators on the savannah to stabilize the newly evolved ecosystem by keeping the numbers of grazers and browsers in check. Klein and Edgar put it well in pointing out,

> In extending their anatomy with tools so that they could behave more like carnivores, they set in train a coevolutionary interaction between brain and behavior that culminated in the modern human ability to adapt to a remarkable range of conditions by culture alone.[2]

This was the beginning of the uniquely human process of evolution in which it can be said, "We invent the technology and then the technology reinvents us."

The dryer climate that converted the margins of the tropical rain forest to a dryer savannah opened a new frontier for the surviving chimps by allowing them to evolve through random mutations into hominids. This new ecosystem, while it created unknown dangers, also released the species from some of the constraints of the rain forest. For example, once they had the tools to hunt, they added more protein to their diet. Modern chimps occasionally hunt, but they have fewer opportunities. But such hunting could not depend on natural weapons such as fangs and claws as the hominids had gradually lost both. But these early hominids could use their uprightness and opposable thumbs to craft their own weapons, possibly creating the first technological innovations. By innovating, they artificially amplified their anatomical fitness. Crude as they were, these first innovations set in motion a process that would later virtually take over the evolution for later human ancestors. Cultural coevolution had begun. But it began very slowly indeed. The first artificial weapons were not even crafted, properly so called. They were likely just opportunistic, broken stones of a size that had a sharp edge and could cut through skin. Similar stones could also be thrown, either in defense or to secure game, because these hominids had throwing arms much like ours but likely with more power. Physically, chimps are much stronger than humans. In other words, fitness for the risen ape, mostly of the species *Australopithecus Africans*, soon became calibrated to the savannah ecosystem. It then altered very little for the next three million years.

During this second climatic change, the savannah became dryer yet and contained fewer trees, had more grass, more predators, and more prey. These dryer conditions apparently created more opportunities, and that once again allowed for new mutations or innovations in human DNA. These new mutations favored an enlarged brain on the upright hominid body, a brain about double in size to about 700 or 800 cubic centimeters. We don't know just when the brain began to enlarge, but it was more or less in place by about one and a half mya. Paleontologists call this human-like creature *Homo ergaster* and in its Asian version, *Homo erectus* (see Figure 2). The most obvious sign of greater intelligence was a quick improvement in the Oldowon tool kit with a wider suite of tools and weapons as illustrated in the next chapter. While still crude, these tools had clearly been crafted. We can see how our ancestors' ability to stand upright and run and to use an opposable thumb favored genetic mutations that enlarged the brain.

When that brain doubled in size, the Oldowon tool kit improved and that fact in turn would later favor a bigger brain yet. However, it was the dryer climate that created a new frontier both of opportunity and danger, and which favored mutations leading to a bigger brain that among other things led to a smarter animal. That mutation would provide enough extra fitness to propagate quickly throughout a small population.

But why would not a bigger brain find favor among the earlier hominids, before the second major dry spell? The answer is about the same as with the upright stance in the tropical rain forest, that is to say, a bigger brain might have come at too high a cost in the earlier ecosystem. The larger brain case had to pass through the birth canal and its expanded size made that passage more difficult. As a practical matter, a big brain case required an infant to be born "premature" and thus much more

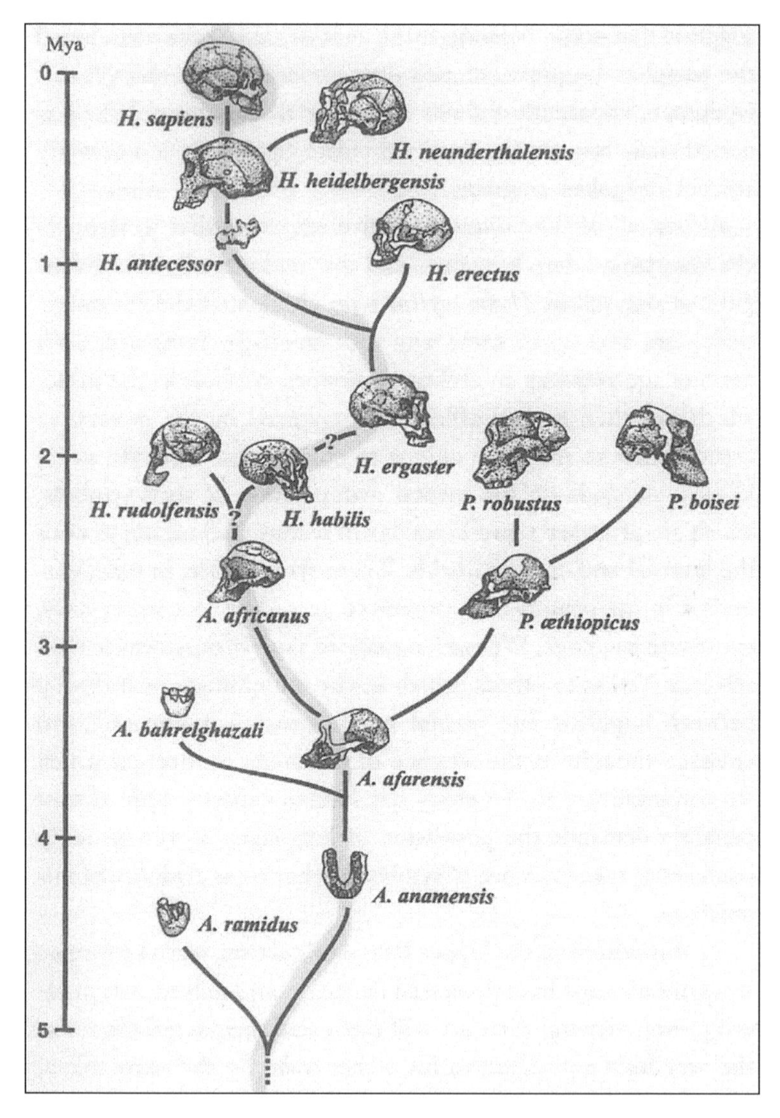

Figure 2: One possible scheme of relations among the various known species of the hominid family. *The known fossil record is simply the tip of the iceberg, and the true picture is certainly much more complex than this. Illustration from* Becoming Human, *by Ian Tattersal (reprinted with permission).*

helpless than before. Most animals have at least some degree of self-sufficiency at birth. Big brained humans have almost none and that fact puts a greater burden on the parents in caring for the infant. That is a high price to pay for getting smarter and it would likely convey no net additional fitness unless being smarter paid truly great dividends. But such dividends would come only if the mutation addressed the new dangers and opportunities of the newly disrupted ecosystem. Moreover, they also probably required some mutations within the brain that encouraged greater cooperation and sharing among bigger brained parents in order to care for their infants. Among other things males would have to want to take a more active role in such caring, behavior that is rare among other primates.

3

THE ICE AGES AND A BIGGER BRAIN

Shifting tectonic plates caused the once-separate continents of North and South America to collide at the Isthmus of Panama almost three mya.[1] According to neurobiologist William Calvin and others, that collision may have helped set the stage for a series of Ice Ages that began soon thereafter, because it precluded the previous free circulation and exchange of waters between the Atlantic and Pacific oceans. In effect, it disrupted the exchange of salt-heavy water coming back from the north end of the Gulf Stream to return to the Pacific.[2] When the two continents came together, water exchange could take place only in the relatively narrow passage between the tip of South America and Antarctica. The closure further disrupted ocean currents around the globe, because the exchange of warm and cold waters between the Arctic and Tropic regions slowed sharply.

According to some estimates, based on core samples of ice from Greenland and other frozen places, along with other geologic evidence, these Ice Ages have come and gone seventeen times since this series began.[3] The hominid line began branching out about this time. Besides *Homo ergaster*, fossils have been discovered from a hominid called *Homo habilis*, which may be modern humans' direct ancestor. Another hominid named *Homo rudolfensis* apparently was unable to adapt and went extinct shortly thereafter. Our *Homo* line, according to current estimates, evolved from *Homo ergaster* almost two mya to *Homo heidelbergensis* about five hundred thousand years ago (kya) and onto our more direct ancestor, *Homo sapiens,* about fifty kya. *Homo neanderthalensis* (Neanderthal) was apparently not in our direct line and also failed to adapt to a changing world. Neanderthals became extinct about thirty kya. So did the Asian branches of *Homo erectus* (refer again to Figure 2). Meanwhile, it seems clear that truly modern people, judged by DNA samples, first evolved in tropical Africa long after the first migrations of *Homo erectus* from Africa to East Asia. Paul Ehrlich states that, "Gradually, the weight of scientific opinion has shifted toward some version of the 'out of Africa' scenario, which has become the dominant view among human geneticists."[4]

There have been many fine books recounting the fascinating story of the search for human origins in the fossil records and including the story of the paleontologists, both amateur and professional, who discovered them, as well as the many controversies that arose in the course of their adventures. Raymond Dart, Louis and Mary Leaky and their son Richard, Donald Johansson, Brian Edgar and many, many, others have all made major contributions to this saga. For my purposes, I simply build on that work and take the current mainstream view as depicted in Figure 2 to set the stage for my own primary interest, which mainly concerns the evolution of human culture *after* it began to trump the evolution of our anatomy. There are, of course, questions, and Paul Ehrlich has mentioned some of them. Why did the first stone-tool culture last so long? Why did it then undergo a couple of hundred thousand years of gradual transition and then suddenly blossom into a culture that so rapidly, in a few tens of millennia led to the rise of nuclear weapons and supercomputers?"[5]

I hold with those scientists such as William Calvin who believe the Ice Ages provided the kind of frontier stimulus needed for this transition. It seems logical to presume that the many disruptions to the many ecosystems of the many Ice Age shifts from cold to warm and back again provided a series of new frontiers. Had the Ice Ages not come, the human race might well have gone into stasis with *Homo ergaster* or *Homo erectus*. If so, no culture would have evolved to the point that someone could write a book about it. But another issue has also been raised about the role of language. As Ehrlich says,

> In other words did the physical evolution of our ancestors' brains cause the Great Leap Forward—or did only the 'software' of culture change, not the 'wetware' of brain structure? And what, if any, was the role played by the evolution of language in that cultural Great Leap?[6]

One school of thought suggests that our modern abilities for language arose quite late in our evolution and account for the Great Leap Forward. Left unanswered is why modern language abilities took so much time to evolve so long after the human brain reached its present size? But if it did take that much time, then how could modern language ability spread so quickly? After all, by then the human species had spread over most of the globe, from its origins in Africa to Asia and Northern Europe all the way to Tasmania off the south coast of Australia.

Still, in appearance we have not much changed for the last two hundred and fifty thousand years, but our tools kept getting better. Figure 3 shows the improvements from Oldowon to the early Acheulean tool kits just after the Ice Ages began.

It is difficult to flesh out the FROCA process in detail during the million-plus-year period of oscillating Ice Age we know as the Pleistocene. But we know humans did adapt to the cold and they would have been put through many shifts in the climate from cold to hot with cycles of about one hundred and thirty thousand years. We also know that before the last Ice Age began to warm about fifteen thousand years ago, our Great Leap forward in culture had already taken place. From core samples we know that the icy parts of the Ice Ages lasted much longer than the warm parts as Figure 4 shows. The warm periods last only about fifteen thousand years as we can see from the chart. According to Calvin and others, our present warm period may be close to its end despite recent trends in global warming.

Some branches of the *Homo* line may have gone extinct if they were caught out by these cold periods. We have a historical

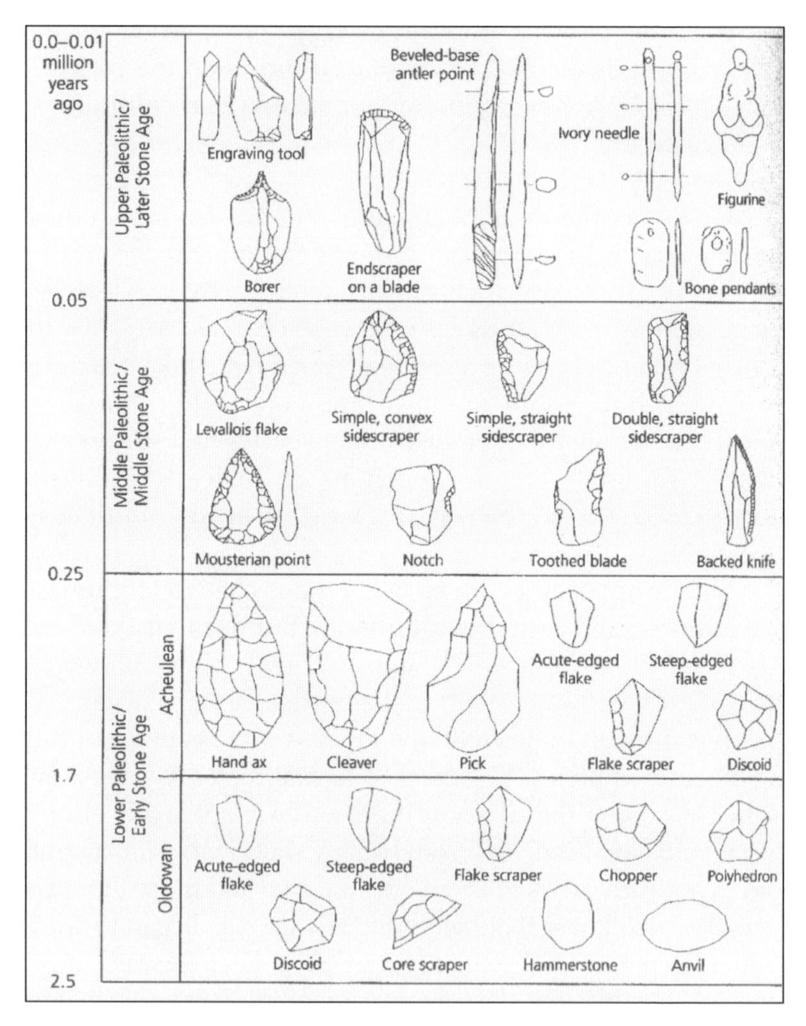

Figure 3: Tools of prehistoric peoples. *Note that the divisions of the time scale are not of equal length. Modified and redrawn from* The Dawn of Human Culture *(reprinted with permission from Paul Ehrlich).*

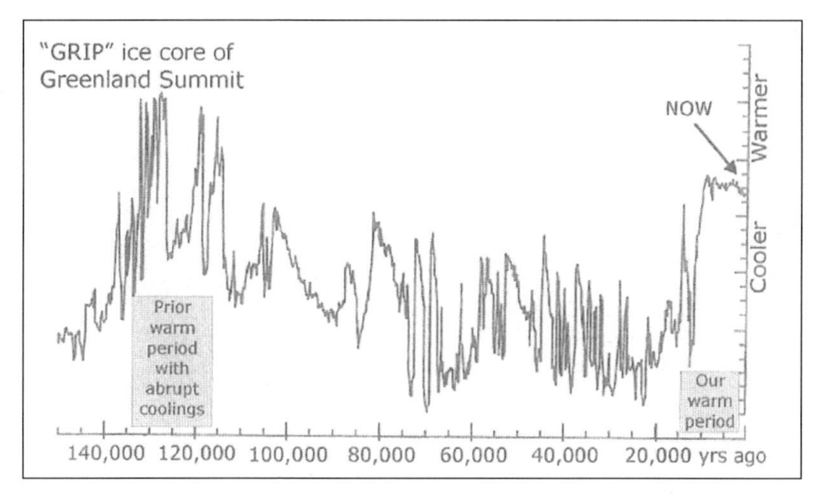

Figure 4: "GRIP" ice core of Greenland summit, *from* A Brain for All Seasons, *William H. Calvin (reprinted with permission).*

example of such an event. Greenland was settled by Norse colonists about 800 CE during a relatively warm period that permitted farming. But then the climate began to cool into what we call the "little ice age." By 1500 CE, well into the Little Ice Age, the Norse colony had vanished. The Inuit had moved back in. Jared Diamond points out that Greenland is the only known historical example where hunter-gatherers displaced farmers. [7]

While technology in the form of better tools and weapons was interacting with our evolving larger brain to increase the sophistication of human culture, the first really major impact did not arrive until about sixty thousand years ago. It was at this point that techno-culture began to come into its own and make major transformations in the way we live. Humans had migrated all the way to Australia and Tasmania by at least forty thousand years ago, or twenty-five thousand years before the last Ice Age began to melt. It was at this time the famous cave paintings and other forms of art became common in Europe. And it was a result of these new artifacts of human techno-culture that humans were led out of the Neolithic era, otherwise called the Stone Age. One of the crucial series of new inventions was a variety of long distance weapons such as the spear thrower, bow and arrow, and boomerang. Humans also developed the tools for much more sophisticated fishing, including boats. These innovations much improved the productivity of the hunt and allowed humans to hunt larger and more dangerous animals. They could also better fend off and drive away other predators hunting the same game. These ancestors developed the first non-agricultural settlements, usually at sites with abundant fish and other seafood. Humans were able to penetrate and live in arctic climes for the first time.

In some areas, such as the Middle East, these new weapons turned humans into much more deadly predators and put them on a much faster FROCA track. Long-distance weapons caused over-hunting, the downside of humans as the globe's master predators. Still, game was wiped out in a certain area, as happened time and again, humans disrupted the local ecosystem. Indeed, efficient hunting led to the first documented major crash, but in single localities at a time. In those localities, humans had to abandon hunting and turn to agriculture, and then later pioneer a whole new life style. This shift from hunting to farming would launch the most dramatic cultural evolution of that time.

4

TECHNO-CULTURE TAKES OVER

The Ice Ages seem likely to have provided the disruptions to stable ecosystems that opened up new frontiers of both risk and opportunity. These frontiers then favored the cultural innovations in technology that allowed early humans to reduce the risks and exploit these opportunities. The innovations picked up speed as time passed and before long they had become the primary drivers of human evolution.

As early as fifty thousand years ago, but certainly by about thirty thousand years ago, crafted human innovations had amplified our physical anatomy in so many ways that techno-culture effectively took over from genetic mutations. Certainly after we acquired the physical and mental abilities to speak in languages, culture became the driving force for nearly all change. Despite our tropical origins, we needed no mutations to allow us to live in cold climes. As we had lost nearly all our body hair, the cooler climate induced us to invent the tools to make clothing to keep us warm. We first learned how to use fire to keep us warm at night, and we learned to make fire when we needed it. Later, we learned how to use fire to extend the practical range of our diet. Cooking changes some foods such as grain to be more digestible and without cooking we could not eat them. Thus, the ability to cook made genetic mutations to allow us to digest raw grains largely pointless. We invented a wide variety of weapons to defend ourselves from stronger, larger, and faster predators. We thereafter had no need for fangs, claws, or horns. By inventing boats, first with paddles or oars and later with sails, we learned to cross wide bodies of water.

The Polynesians indeed managed to cross nearly the whole of the Pacific Ocean. None of that seems to have required any genetic improvement in our ability to swim, and humans remain far behind the fish and amphibians in that skill.

Many of these innovations took place during the cold period of the last Ice Age that began to warm up about fifteen thousand years ago. Up to this period, most estimates put the total human population at between five and seven million people over the whole globe shortly before the population explosion that was to come.

The sudden warming trend itself opened a new frontier in North America that led to its first colonization (at least on a large scale) by modern humans. At the same time, this warming trend opened up a new way of life for those hunter–gatherers in what is often called the Fertile Crescent. This area of land goes roughly from the Persian Gulf up the Tigris and Euphrates river systems of Iraq, into Turkey and Syria, and then down along the Mediterranean Sea via Lebanon, Israel, and Palestine to the Red Sea. Nomadic hunting and gathering gave way in a relatively short time to settled agriculture. Many archeologists suspect this transition was not voluntary. It may well have been forced by overexploiting our dietary prey.

Before we examine the first transition to agriculture, let us first look at the colonization of North America. It led to one of the most startling examples of overexploitation that we know. Similar kill-offs took place along the Pacific Rim including Australia, New Zealand, Hawaii, and Easter Island. It also happened in Madagascar and elsewhere. Here, I follow Tim Flannery's account as recounted in his comprehensive study of the ecology of North America since the Yucatan asteroid hit of sixty-five mya.[1]

Just before the last heavy glaciations ended about fifteen thousand years ago, North America contained the largest population of mega fauna on earth, mostly large mammals. Giant sloths, woolly mammoths, mastodons, elephants, giant bears, saber-toothed tigers, and many others abounded. Yet in hardly more than three hundred years, nearly all had disappeared and during a period of rapid warming. At one time the accepted view was that this mass extinction was a function of climate. Later analysis cast grave doubt on that view. The newer view is that the Clovis people, the first human immigrants into North America (south of Alaska) did the job. The story as told by Flannery and others is that once the Clovis people broke through the ice somewhere around Edmonton, Alberta, they found themselves on a new frontier of unparalleled opportunity.

Unlike the large mammals of Africa and Eurasia, those in North America had never confronted human hunters. Thus they did not fear them. Humans were small and hence did not appear threatening. The Clovis people found the hunting unbelievably rich and easy. It was true ecological release on a new frontier awash in opportunity. Indeed, to take advantage of those opportunities they invented a new inventory of weapons, suited to the cornucopia of game. No Clovis weapons have been found anywhere outside North America. They were not the weapons the Clovis people brought with them from Eurasia, which were much smaller, as were those from most other Eurasian hunter–gatherer cultures. Tim Flannery puts it this way:

> As these people entered their New World their toolkit underwent an astonishing transformation. Almost instantly they began manufacturing an entirely new kind of instrument, the Clovis point. These magnificent fluted stone spearheads are of distinctive design, ranging from around four to twenty three centimeters in length. [2]

Flannery goes on to point out that the Clovis people made these extraordinary weapons (as determined by radio carbon dating) for a mere three hundred years. Why? Because, he says, by about twelve thousand nine hundred years ago, these weapons had been used to kill off all very large creatures of America and thereafter the Clovis points were too large to be useful against the smaller (and now wiser) remaining game.

In North America, the mega fauna of North America had survived seventeen consecutive cycles of glaciations followed by warming, and survived just fine. Moreover, the one place where Mega fauna did survive for a longer period was on the island of Cuba until about six thousand years ago and then vanished just after the first humans arrived about that time.[3] In every case, and in a variety of climatic circumstances, the mega fauna vanished shortly after the humans arrived in the area.

As mentioned, this hunter–gatherer kill-off in North America had counterparts elsewhere. It happened much earlier in Australia at the hands of the Aborigines about thirty thousand years ago. All mega fauna larger than human beings were wiped out, including very large marsupials. Flannery explains, "Australia lost every terrestrial vertebrate species larger than a human as well as many smaller mammals. . . . In all, about sixty vertebrate species were lost, including bizarre marsupials that resembled giant sloths (and others)."[4] It happened much later in New Zealand with the arrival of Maori about 1200 CE. In New Zealand, eleven species of a large flightless bird called the moa were wiped out within about two hundred years after Maori arrived.[5] The arrival of humans in nearly all the islands settled by Polynesians in the Pacific, including Hawaii, proved devastating to preexisting ecosystems.

One reason many people resisted the idea of human causes for disappearance of the mega fauna was a romantic notion. It was widely believe that primitive peoples, Amerindians in particular, were ecologically minded and took great care of their environment with which they were spiritually in tune. Chief Seattle's prayer is often quoted: in part it says, "Every part of the earth is sacred to my people. Every shining pine needle, every sandy shore, every mist in the dark woods, every meadow, every humming insect. All are holy in the memory and experience of my people"

I was taught that trashing the environment was a crime perpetrated mainly by white settlers from Europe influenced mainly by their industrial culture driven by capitalism's profit motive. While the industrial revolution clearly produced a great deal of industrial pollution and other devastation of the natural ecosystem, we now have clear evidence such devastation happened regardless of whether the industrial culture's software is capitalist, communist, or mixed. We know that because when the Cold War ended, we soon learned that the communists did far more environmental damage to their ecosystems than America did to its own. In fact, the post-communist regime in the late Soviet Union on occasion asked for our help to remediate the damage they had done.

I do not claim Chief Seattle's prayer was false or hypocritical. Rather, I suggest that the Amerindians probably developed a greater spiritual awareness of their environment as an adaptation following the Crash of the Great-Mega-Fauna-Kill-off. Like many who participated in the ecological release of the recent dot.com boom, these early immigrants to a new land found they also had to make painful adjustments following the bust. Those early Clovis peoples had grown used to an extraordinarily high standard of nomadic living thanks to their hunting tools and the huge native herds of mega fauna. Moreover, the Clovis people spread out quickly over the continent. Then, like the collapse of the dot.coms, they ran out of game. They were forced to make painful adjustments to a new ecosystem where they could no longer "get rich quick" with easy hunting. Part of their adjustment may well have been to take more account of and greater care to conserve the natural environment. Introducing the cultural concept that the game they preyed upon was sacred would help preserve stability and equilibrium in their post-crash adjustment.

The Clovis people's migration in North America created almost no regional cultural variation, judging by remaining artifacts, according to Flannery. They apparently remained intently focused on exploiting their opportunity to hunt big game to the point of its extinction. In this they acted little differently than did Euro-American pioneers and settlers in the nineteenth century that nearly wiped out the American bison and did wipe out the passenger pigeons whose population was once measured in billions. What the shotgun did to the passenger pigeon and the repeating rifle nearly did to the bison, the Clovis points did to North America's mega fauna of thirteen thousand years ago.

But after the crash took place, the single Clovis culture, together with its tools, abruptly vanished and in its place a wide variety of local cultures emerged and adapted to the local conditions: geography, climate, flora, and fauna, but now without the mega fauna that had maintained Clovis homogeneity. Flannery notes, "Such cultural homogeneity has never again been seen in North America."[6]

One could argue that a similar environmental awakening is underway among industrial cultures today. Moreover, once it was clear we had wiped out the passenger pigeons and killed nearly all bison, Americans too had second thoughts. America's conservation movement dates from that time, an early step perhaps in a new adaptation.

The Clovis peoples exhibited an almost classic example of the FROCA process among primitives little different in principle from the dot.com boom and bust thirteen thousand years later. A new frontier opens up, new risks and opportunities appear, and the pioneers at once began to exploit them. A new technology quickly evolves to take advantage of those opportunities and often to create new ones as well. The opportunities are overexploited and trigger a crash, followed by a new adaptation.

Occasionally, however, these adaptations do not go well. The story of Easter Island, as recounted by Paul Ehrlich, provides a good example.[7] About sixty or so Polynesians first colonized the island around 400 CE. They brought with them pigs, chickens, and rats (by accident, most likely) along with breadfruit and other fruit trees. Only the chickens and the rats survived. Still, the pioneers found a good diet made up of nesting seabirds, fish, and even dolphins, which they caught from aboard large canoes they made from slow-growing local palms that covered most of the island's sixty square miles. On this new frontier, the islanders enjoyed a whole host of new opportunities that they began to exploit. By about 1400 CE, their population had grown to about ten thousand people with a rich culture symbolized by the huge stone moai (statues) that made Easter Island famous. Some statues weighed more than eighty tons and stood twenty feet high.

However, the crash soon began. One source of trouble was the megaliths. The islanders had erected their stone megaliths by using log rollers. They cut the logs from the forests of slow-growing palms and then used those logs to move the stone that they quarried from mines several miles away from the erection sites. The islanders also had divided into at least two competitive groups. Both groups erected their own megaliths. Neither side would quit the competition until they had cut down every last palm tree. Not only did they run out of log rollers, they also had no more wood with which to build their canoes used to catch fish and dolphins. The seabirds also vanished with the palm trees. The lack of forests caused the soil to erode so that growing vegetables became ever more difficult. As Ehrlich has described it, "With the forests went much of the soil, swept away by strong drying winds in the absence of windbreaks. Crops doubtless declined, and warfare over dwindling resources became common."[8] Ehrlich goes on to point out that the oral histories revealing this also revealed the practice of cannibalism. A classic taunt became common, "The flesh of your mother sticks between my teeth."[9] By the time Europeans arrived in the eighteenth century, the population had crashed to about two thousand and continued to decline until falling to its low in 1900 of 111 survivors.

Coming back to the Clovis peoples, as the process of local adaptation evolved, North American hunter–gatherers developed scores of different cultures, customs, and languages, along with smaller spear points and arrow heads more suited to the smaller game. They thus fragmented into the sort of tribalism typical of most advanced hunter–gatherer societies. Like the

Easter Islanders, Amerindian tribes fought each other on a more or less on-going basis. But, with much more territory at hand, the results were far less devastating. One could even argue that, given ample territory, low-grade skirmishing kept the tribes from aggregating to the point where they could not sustain their populations on the local resource base.

Moreover, once the mega fauna vanished, local ecosystems also changed as smaller animals now became more prominent and human hunters added another dimension to the competitive interaction and cooperation. For example, as hunters used fire to drive some herd animals off cliffs or into bogs, the local flora became more fire resistant. Indeed, that adjustment was far more prominent in Africa than in America, given the evidence of fossil plant remains, according to Jonathan Kingdon.[10]

Not until humans turned to farming to supplement their diet because of dwindling game animals did permanent settlements arise on a large scale. But there were some exceptions mostly in areas rich in seafood. One such area was the Puget Sound region of Washington State and British Columbia up through the Inland Passage to Alaska. Rich in salmon and halibut as well as clams, oysters, crab, and even whales, the Amerindian tribes were able to settle in permanent villages. Of course, they first had to invent the large canoes and fishing gear to make such a way of life possible. This region was heavily forested on hills and mountains making traditional hunting problematic compared to plains, prairies, or savannahs. But these early settlements and more predictable lifestyle allowed for a big increase in artistic expression as represented by totem poles and other carvings. These villages also encouraged certain forms of behavior that was later to become far more characteristic of humans once they turned to farming. For example, staying in one place allowed people to accumulate wealth of sorts in such rich fishing grounds. Hunter–gatherers, however, can almost never accumulate wealth because they must be constantly on the move to find more meat, and in any event the meat won't keep for very long. It can be cached and frozen in the winter in some places. Moreover, nomadic hunter–gatherers had to carry everything with them as they moved before they domesticated animals or invented the wheel.

Still, the Indians of the Puget Sound area could set traps for fish in the streams and tidal reaches so that the fish came to them. For those who had to fish in the open water, they could go to the fish by boat, and then return home to settled villages and their long houses. They could smoke the fish and it would keep for considerable periods. Moreover, with boats they could carry more artifacts if they had to move the village. As such wealth accumulated so did competition for it. It then became necessary to establish some form of "law and order." Village headmen, then chiefs, and later chieftains evolved with increasing authority. They had the power to resolve disputes among members in now fairly populous villages. Another important difference compared to pure hunter–gatherers was that greater wealth, based on the rich fishing resources, allowed for division of labor. Not everyone had to fish. Some people could make tools, boats, harpoons, nets, and totem poles. The religion of shamanism came into being as did a developed spiritual life based on the totem poles. Moreover, and this was another sharp departure from the hunter–gatherer lifestyle, richer villages began to acquire slaves, usually the captives of tribal warfare.

Sometimes the competition between villages took the form of a potlatch rather than a pitched battle. The potlatch was a form of party in which village chiefs would compete with one another to see who could throw the biggest bash and give away the most wealth in the form of elaborate feasts.[11]

Let us be clear, however, that such cultural elaboration was clearly dependent on the technology that made fishing in its various forms possible, combined of course with the rich fishing grounds. The Amerindian tribes of the Pacific Northwest could not over fish as the fishing was too rich for the relatively small population. Not until European settlers arrived with huge increases in population did overexploitation of fishing begin on a large scale with the added technology of deep sea fishing.

These early pre-agricultural Stone Age settlements prefigured behavior that would soon come as humans began, in one locale after another, to turn to farming. This transition to the production, rather than simply gathering, of food began in the hills of the Fertile Crescent. Farming later spread to the river valleys as population grew, first in the Tigris and Euphrates and later in various other river valleys such as the Nile, Indus, Ganges, Yellow, and Yangtze. A huge population explosion was soon to follow, and to that story we turn next.

PART II

EARLY FARMING TO DESPOTIC CIVILIZATION

5

FARMING AS A NEW FRONTIER

Humans began serious and systematic farming in the Fertile Crescent of the Middle East about ten thousand years ago. The transition from hunting and gathering to farming was both an adaptation to the decline of hunting and the start of a new frontier of sorts that emerged from the older gathering tradition.

The transition, however, was both gradual and localized. Where farming began, several generations followed to complete the transition in the local area. The agricultural frontier and subsequent Cultural Revolution took about four thousand years to unfold from local garden farming to sophisticated high output agriculture. During that time, a wide range of new technology emerged that made sustainable civilization possible.

However, human anatomy and psychology had been calibrated to nomadic hunting and gathering for up to seven million years. So during most of the transition to farming, men continued to hunt but on a slowly reducing scale. People continued to gather wild plant food, but again on a reducing scale. Still, the longer one group stayed in one place, the more quickly they reduced populations of local edible prey and plants, and the more were forced to depend on learning how to cultivate plants in a settled location. One widely held view is

> that the domestication of plants was the work of women in the hills of Northern Iraq or Kurdistan. The women, left behind in temporary camps while their men hunted sheep and goats in the nearby hills, harvested wild seeds and eventually began to cultivate

> them. . . . The entire basis of existence was thereby revolutionized with consequences that are very much a part of our lives in the twenty-first century.[1]

Gathering had long improved human knowledge about the life cycle of plants and how to cultivate them. For example, people would have noticed that upon returning to some longer lasting campsite, the plants near latrine areas had the most vigorous growth. It was thus a short evolutionary step in agriculture to use human or animal manure as a fertilizer. Indeed, the use of animal manure as fertilizer (and in some cases as fuel) may have been an early motive for the domestication of animals. They were also a source of meat or milk and could be used for work, such as pulling a plow or hauling material.[2] In addition to fertilizer, people began to invent better instruments of cultivation from mere sticks to break the ground to tools such as the hoe, and later the plow, for that purpose.

Jared Diamond points out that considerable experience and insight was needed to select the proper plants for cultivation and the proper animals for domestication. As he puts it, "Among wild plant and animal species only a small minority is edible for humans or worth hunting and gathering."[3] They may be indigestible, poisonous, tedious to prepare as with very small nuts of very low nutritional value, or too dangerous to hunt. Still, after the transition, a given territory might supply more than one hundred times as much food as the equivalent hunter–gatherer yield in the wild.

The domestication of animals counts as a major innovation in its own right. In fact, it helped determine that Eurasians would come to dominate the globe. As we will see in later detail, the absence of animals that lent themselves to domestication conveyed a crucial competitive disadvantage to the peoples of Australia, as well as the native peoples of North and South America compared to Eurasians. As we shall shortly see, the absence of domesticated animals was to assure such peoples an absence of immunity to diseases to which Eurasians had long become accommodated.

Additionally, as Jared Diamond puts it,

> Before animal domestication, the sole means of transporting goods and people by land was on the backs of humans. Large mammals changed that: for the first time in human history, it became possible to move heavy goods in large quantities, as well as people, rapidly overland for long distances."[4]

Thus the innovation of animal domestication opened new frontiers of opportunity in the form of mobility for whoever learned how to use them. The horse was to prove one of the most important of these and became "the Sherman tank" of the ancient military world until the early part of the twentieth century. The horse was first domesticated on the steppes of Russia or the Ukraine north of the Black Sea. Those with horses conquered those without them and helped spread the conqueror's languages among the conquered.[5]

Meanwhile, the domestication of animals enabled the Europeans after 1500 CE to quickly subdue the Americas and Australia quite apart from the military power of the horse. The reason concerns crowd diseases such as smallpox. Many or even most of these viral diseases began with animals such as cows, goats, sheep, pigs, horses and others. Domesticated animals lived in close contact with each other and with their human minders who also lived in close contact with each other in settled villages. Thus, when an occasional mutant form of a viral disease arose in some animal that enabled the virus to thrive in a human body, the disease would first spread among the animals and then jump to the now more densely settled humans. Such diseases were soon a major cause of human mortality and a major side effect of domesticating animals. Still, that price appeared to be worth paying because those same animals helped transform human existence.

Crowd diseases evolved by the same FROCA process that applies to all other biological and cultural forms of evolution. As an example, Spanish explorers and conquistadores unwittingly brought smallpox to Central America with the arrival of Columbus. In the bodies of the Spaniards, the smallpox virus lived under serious ecosystem constraints because the Spaniards had long ago developed a high degree of immunity to the virus. Although the disease became active occasionally, when a person's immune system grew weak, smallpox was not usually fatal to Spaniards. But, upon their arrival in the New World, close contact including sexual encounters with the Amerindians allowed some of the Spanish smallpox viruses to migrate. The bodies of the Amerindians became a brand new frontier of opportunity for growth as far as the virus was concerned because those bodies had no immunity at all to smallpox. The virus at once began an explosive growth, jumping from one body to another without constraint. Of the estimated twenty million Amerindians of Mexico, more than 90 percent died of the disease within fifty years. Similar plagues broke out in Peru and later in other parts of the Americas. The process was repeated in Australia with the arrival of the English settlers in the eighteenth century.

Later, however, the viral expansion crashed. As the disease was almost always fatal at first, the disease had soon overexploited its "opportunities." When host communities died off, so did the smallpox viruses. Before too long the most vulnerable humans were dead and so a huge drop took place in the viral population. Adapting to equilibrium, the smallpox virus lived only within the ecosystems that constrained them with immune factors. A few of the Amerindian victims of the disease had acquired mutations that created the various immune factors needed to constrain smallpox. They survived the disease and passed on their immunity within the diminished Amerindian population. From these few mutant survivors, the Amerindian population began to recover and acquire a new immunity to the disease comparable to that of the Spaniards.

But by the time the native population recovered, the Europeans had taken control of both American continents from top to bottom as would later happen in Australia.

The reason the Amerindians had no crowd disease immunity simply reflects the fact that, except for dogs and, in the Andes mountains, the llama, neither American continent had animals that lent themselves to domestication. Some that might have served this purpose had been killed off twelve thousand years earlier. Was the smallpox plague a revenge of the extinct mega fauna on the Amerindian ancestors of the Clovis peoples? There were of course other immigrations into the Americas from Asia after the Clovis people entered and before the Europeans arrived, but there was much intermingling thereafter.

The spread of farming as a new human frontier was heavily influenced by factors of geography and climate.[6] It has, of course, always been obvious that well-watered areas located in the temperate zone were ideal for raising many crops. A temperate desert or arctic tundra won't do. The hot temperatures and heavy rainfall of many tropical areas also leach nutrients from the soil. But the temperate parts of the Eurasian land mass heavily favored the cultural spread of farming because of its east–west axis of orientation along given latitudes.

Farming spread from its origins in the Fertile Crescent most rapidly along that east–west latitude (between 40 and 50 degrees north) for several technical reasons. First, the latitude of an area most influences its climate despite wide variations in other factors such as rainfall. For example, the proportion of day and night during seasonal cycles stays the same along any given latitude. Thus, the planting, growing, and harvesting seasons will be more or less the same along a latitude.

The same cannot be said going significant distances north and south. When the seed is ready to sprout, for instance, in Southern Oregon, snow may still be on the ground in North Dakota. Growing seasons vary sharply north to south but not east to west. Thus, a variety of wheat or corn, for example, can spread more quickly east to west. Years of cross-breeding were required for most grains before new varieties evolved that could thrive in a variety of growing seasons apart from that of the original wild plant. The north–south orientation of the Americas largely accounts for the slow spread of agriculture there compared to Eurasia. Corn did spread north of Mexico, but not until appropriate varieties had been cross-bred that could handle the colder seasons to the north.

Although rainfall an vary considerably along any given latitude, that variation did not constrain the spread of farming east to west nearly as much as the differences in growing seasons constrained it north to south. Geography made this difference. Originally, farming spread from the well-watered hills of Anatolia into those hot and dry areas that lay along large river valleys. The first such valley lay close by, namely the Tigris and Euphrates valley of modern Iraq (or ancient Babylon). Here, farming became more extensive because it was nourished anew each year by the silt brought down from the mountains by its Twin Rivers.

Farming then migrated west to the Nile valley of Egypt, and east to the Indus and Ganges valleys of India, Pakistan, and Bangladesh. It evolved independently in China along the Yangtze and Yellow Rivers somewhat later. It also evolved independently and separately in Mexico and in the Andes mountains of Peru. Again, in all such areas, farming migrated east to west much faster than north to south. The only continent aside from Antarctica that never developed agriculture was Australia. It remained firmly a land of hunter–gatherers until the British arrived in the eighteenth century.

Technological advancement in the emerging agricultural societies depended heavily on what kinds of animals could be domesticated. In the case of North and South America, there were almost none. In most other areas, various species of cows, horses, pigs, goats, and sheep were common. Chickens, ducks, and geese were the most common species of fowl. The larger mammals in particular opened the door to crucial innovations that later civilizations would require, such as the wheel. From the wheel came variations of it, first perhaps in wagons, carts, chariots, but also in other tools such as pumps, pulleys, windlasses, rope, and better metal crafting tools made out of copper and then bronze.

By today's standards, the agricultural way of life spread rather slowly. It emerged from the hunter–gatherers and then, as it spread, displaced them. From about 8,000 BCE to 4,000 BCE, Eurasian agriculture encompassed more and more area to become the standard way humans fed themselves. Thus, humans invented more and more tools, weapons, and other instruments that were added to the inventory of human cultural technology and became incorporated into the cultural technostructure. Compared to previous rates of evolution, this new technology came as a sudden spurt. It was agriculture, including the domestication of animals, that made this new technology possible.

The reason is three-fold. First, and unlike the nomadic hunter–gatherer communities, farming produced a surplus of food over and above the narrow needs of the immediate food producers and their dependents. Life was no longer on such a short day-to-day subsistence leash. Moreover, farm children were an asset, not a liability. They could perform many tasks of farming such as weeding crops or minding some of the animals. Children in a hunter-gathering economy were often a burden to nomads who had to carry everything they owned wherever they went. Second, surplus food production freed people to make and occasionally invent the specialized tools to fit particular circumstances. The division of labor emerged as some people began to specialize in making tools, or else in using them in ways aside from direct food production. Nomadic hunter–gatherers could not specialize. Third, the settled way of life, division of labor, and surplus food all together freed up a greater amount of time for reflection and experimentation and the ability to learn through trial and error.

Moreover, one new thing would lead to another as we have already seen. Making animals available to carry goods or pull plows led to the invention of wheels, then to carts, and so on.

In settled communities soft metals such as copper or tin might accidentally be smelted out. Once cooled they could be played with and perhaps fashioned. Before long, the Stone Age was gone, replaced by a new and much more sophisticated Bronze Age. As humans advanced their technology, the same technology began to change human culture. No longer exclusively nomads, people found more time for spiritual reflection, artistic expression, playing games, and telling stories. Indeed, they could now often ride from place to place instead of walk. Moreover, the population exploded in the farming communities compared to the sparsely settled areas still controlled by hunter–gatherers. Tribal villages became more common and larger. Not only were the number of these villages growing as farm technology and know-how spread, many of these villages were growing into towns and soon into cities. This settled life led to a population explosion that we will examine in Chapter 6.

6

RELEASE AND EXPLOITATION: A POPULATION EXPLOSION

The ability to store surplus food, once again, changed our culture forever. Above all, agriculture released humans from a million years of constraint on population growth that was embedded in the hunter–gatherer ecosystem. Our numbers expanded only as we gradually used our technology to extend the range of our hunter–gatherer territory into colder climes.

Birth rates remained low because a nomadic day to day existence without transport sharply constrained the number of children human mothers could raise. Recall that our evolving big brains and a constricted birth canal forced mothers to give birth months before an infant had the least bit of self-sufficiency. For more than a year after birth, human infants would require intensive care all the time by mothers continually on the move in search of food. Thus, almost all hunter–gatherer societies developed strong forms of birth control, such as mothers avoiding pregnancy by nursing their infants, and then nursing them until age four. It was therefore rare for a nomadic mother to give birth to more than two or three infants over her average lifetime of thirty-five years.

With a shift to settled agriculture, those constraints largely vanished. Even very young children could weed, for example, and they did not need such constant watching. Malthusian calculus took hold. (For those interested, the formula is: The integral of the natural log e raised to the power of x will equal a function of U, raised to the power of n.)

Suddenly, in the expanding number of farm communities, women began giving birth to more than two or three children.

Five or six children, and even up to ten or twelve, became more common. Most of these children on reaching maturity farmed new plots of ground instead of their parents' land. The new crowd diseases took their toll on some children with weaker immune systems, but if they died, they failed to reproduce. Thus immunity to these diseases began to build up. The agricultural population began to grow at an exponential rate. Before long, the well-watered hill sites where the farming started began to become overcrowded and people moved away and settled in the river valleys.

These valleys, with their annual inundations of silt, proved even more fertile and productive than did the hill sites. What is more, the flat terrain made it possible to build irrigation canals that enlarged the productive areas. These canals proved to be one of the crucial ancient innovations that made civilization possible. They began in Babylon where truly large surpluses of food could be produced in the fertile valleys with ample water flowing down the twin rivers year around. Before long, riverside farmers learned how to dig canals from the river into the dryer land nearby. Large networks of such canals were to become a major achievement of these farming communities and have given rise to the term "hydraulic society." Karl Wittfogel coined that term and he argued that these canals were responsible for the rise of despotism in ancient times. Labor had to be coerced to dig canals that cut across communities because humans had no previous history or experience with intercommunity cooperation.[1] Diamond, however, argues that local despotisms had to precede construction of large scale canals.[2] On reflection I decided that Diamond had the better case and have adopted his view in this book, but will defer further discussion to Chapter 7.

Meanwhile, and for the first time, human population grew large enough to permit the rise of cities. With cities came well-organized political governments and religions, monumental buildings and other public works, tombs, standing armies, and extensive trade routes. In short, these valleys became the cradle of civilization.

This leap by humans into civilizations culturally organized into far more complex societies than before compares to the emergence of the multicelled organisms that began about six hundred million years ago. At that time, individual cells collectively developed into specialized organs that became parts of new and much more complex organisms. They then evolved into the branches of plant and animal life we know today. It would seem the rise of civilization was evolution's first step in taking humans down a similar path of complexity. The sharp cultural division of labor and occupational specialization of emerging civilizations were akin to the earlier specialization of cells into organs such as brains, blood, and bones. Each organ then played a specialized role in the life of the whole organism. Something similar seems to have taken place with the specialization of society made possible by agriculture and its technology. Thus, a new type of social meta-organism began to evolve by about six thousand BCE. This new society was becoming highly specialized but was made up of the same general purpose individuals that evolved in the hunter–gatherer clans. The specialized cells in the organs and tissue of organisms also begins from one general purpose cell.

The driver of all this expansion continued to be what it has always been, the genetic imperative to "go forth and multiply." So multiply we do. Still, one of the main functions of any ecosystem is to constrain expansion via competition when resources become scarce. Scarce resources cannot satisfy the unconstrained demands of all claimants. Competition brings stability and balance into the equation of life and at the same time allows for the biodiversity among the species.

The invention of high-output agriculture followed by the evolution of culture into specialized (and diverse) civilizations gave the human species ecological release from the long-standing constraint on its population from resource competition. However, humans often drove competitors for those resources away, if only unwittingly. At the top of this food chain, humans took what they wanted, and let the others fight over what was left. As agriculture expanded, humans began cross-breeding plant and animal species into "designs" of our own making to serve our own purposes. Few of the foods we now consume, few of the animals we domesticate for food, pets, fiber, or work, are genetically the same as the original wild species. Nearly all of this selective cross-breeding was under firm human control.

Our virtual takeover of our ecosystems via agriculture and its functional technology transformed the broader physical environment in some locales, often for the worse. Partly this was the result of the orders of magnitude increases in population that farming made possible through its higher productivity. Not only did we require more food and fiber, but we made increasing demands for building materials and firewood. Then too our now domesticated herds would often overgraze with damage to the soil. Erosion began and forests began to vanish.

Ultimately the Fertile Crescent became much less fertile and productive as Diamond points out. Still such "collateral damage" was but a foretaste of what would come.

Before we did too much damage, however, we would move on, migrate to newer and more fertile areas, and leave the old behind with fewer people. In other parts of the world, such as in tropical rain forests, agriculture took the form of "slash and burn." First, the trees were cut down and burned, or simply burned down to clear land for planting. Then people formed a settlement in the burned-out areas and began to farm. At first, the land was reasonably fertile because much of the dead biomass from the rain forests remained in place. After clearing, however, the soil was exposed to much more intense heat and heavy direct rainfall, often in the form of tropical deluges. In a few years, the nutrients would be leached out, productivity would collapse, and the village was forced to move to a new site to start the slash and burn cycle again. As long as very few people practiced slash and burn, tropical rainforests recovered, despite the fact that once cleared, such land recovers much more slowly than do temperate forests that can quickly bounce back. If slash and burn populations become too large, they quickly can obliterate a whole rainforest.

While this population expansion was underway, however, enormous social changes also were taking place in places where villages exceeded about two hundred and fifty people. The egalitarian lifestyle that typified hunter–gatherer bands began to collapse, and as it did, a form of murderous anarchy took its place.[3] This anarchy set the stage for the highly specialized civilization via social stratification and slavery imposed by a new force in human affairs—despotic government, a story we take up next.

7

SOCIAL CRASH AS INTRA-TRIBAL ANARCHY

Humans lived and worked in small egalitarian bands for hundreds of thousands of years. Our psychology, including a great capacity for teamwork, cooperation, and sharing, had evolved in that setting; it was calibrated to the hunter–gatherer way of life in its ecosystem. Yet, in a remarkably short period of time that stable social order descended into anarchy. The population explosion in local villages fostered by the shift to agriculture triggered this descent. Earlier we noted that this social shift was foreshadowed by the richly endowed fishing villages that predated agriculture. Because those villagers could abandon the nomadic way of life, they could establish and then settle down in permanent villages and have more offspring. The population promptly began to grow. As it did, the old informal way of settling disputes within small nomadic bands came under stress. When populations grew larger than about sixty people, settling internal disputes informally became more difficult as close bonds between members began to weaken. Moreover, disputes were more likely to arise when greater numbers of people produced food beyond immediate needs and that surplus could be stored for long periods. In this case, questions of distributive fairness begin to surface in a way that rarely happens among nomads who cannot store food. Such questions then intensify disputes within the community in a way that does not happen when everyone shares with everyone else.

In an interview with a British journalist, Isabel Hilton, a Pakistani member of the Pashtun tribe allowed that if conflict arose between him and his brother, it stayed between him and his brother. But if his cousins came inot the picture, the conflict changed to him and his brother teamed against their cousins. Then if villagers from a neighboring tribe came into the picture it was him, his brother and their cousins against those villagers.[1] This brief interview illustrates a long-noted facet of human nature: unity is much easier to achieve among contentious individuals if they agree on a common enemy. If we unpack this point a bit, we can uncover the origins of human despotism.

In the early days of farming the old egalitarian ways prevailed. All production was shared more or less equally throughout the community. Society was held together with a series of mutual obligations and the surplus, if any, was usually small and often used mainly for seed. If the surplus was significantly greater than that, then the community might enjoy a community feast. Everyone participated.[2]

If someone was in need one was expected to give help, and one could also expect help, when needed, from others. But when giving help, one accumulated obligations from others, and when receiving help, one became obligated to others. Obligations had to be repaid, directly or indirectly, if one was to remain a member of the community in good standing. Personal accumulation of wealth was not socially sanctioned; if anyone suddenly acquired a windfall, say from an external source, one was expected to share it out to repay one's accumulated obligations. This ethic survives in a much attenuated form to this day in most families in all cultures but most strongly in those cultures where tribal identities remain important. But even where tribal identities have largely dissipated, as in most of America, a family member who won the Lottery, for example, and did not share a big part of it with other immediate family members, would probably be ostracized, although obligation diminishes for relatives with whom one has no history of mutual sharing.

The problem with this system of "to each according to their needs, and from each according to their ability" is that it works well only in small groups with close ties. Thus when food surpluses accumulated in communities that had grown to two-hundred fifty people, disconnects increasingly arose between what one felt obligated to share and what others felt they deserved. Such disconnects led to the kind of dispute that almost never arose in small, nomadic hunting bands. Even now, the more people who live in a community, the less people feel close to others in that community beyond immediate kin. As the numbers rise, many count as total strangers. So, if a dispute erupted between two people in a large village without formal authority, they may have no mutual friends or family members to help resolve it. Such disputes often led to violence and even murder. If family and friends alone would come to the aid of their kin and support them in the argument, that situation might well become the origin of a family or clan feud. Indeed, we can see something similar in today's "drive-by shootings" by members of rival gangs in some cities.

Growth in the absence of formal government or police led to the collapse of law and order. Anarchy arose within tribes. Factions or family feuds erupted and persist for generations. Diamond watched this process unfold in New Guinea during the transition from hunter–gatherers to agriculture. He noted that the main cause of deaths among adults was murder.[3] As with the Pashtuns whom Isabel Hilton interviewed, it took disputes between tribes to dampen down disputes within tribes, and it took disputes between clans or families to dampen those disputes within families or clans, and so on. Many species of social animals, such as dogs or wolves, display animosity to strangers. So do humans. Such aversion to strangers among small nomadic hunter–gatherer bands probably served a useful purpose in keeping the bands spread out, thus reducing acute competition for resources.

I long thought that "tribalism" applied mainly to pre-modern people or, in this day and age, to the Third World. But as the nineties unfolded, it became clear that tribalism was alive and well. Indeed, it seems to be a basic part of human social psychology that encourages a distinct "us versus them" bias. It provides an incentive both to cooperate and to compete. We band together into cooperative groups to compete against rival groups. Armies, sports teams, and business corporations all draw deeply from this tribal "instinct." Indeed, Japan's business corporations adopted the slogan, "competition between firms, cooperation within them" almost from the outset of their industrial revolution. This slogan well illustrates how competition and cooperation are related parts of a whole instead of a polarity. Each behavior has its place in any ecosystem and tends to reinforce the other.

When the Cold War ended, it quickly became apparent that neither the former Yugoslavia nor the former Soviet Union had transcended tribalism and its tendency to find internal coherence in extended rivalries. Both of these nations had merely repressed the rivalries. The underlying animosities, some going back a thousand years, remained firmly in place. Once Yugoslavia and the Soviet Union separated, old tribal, ethic, or religious animosities erupted into action, a sort of ecological release from the previous constraints. Violence broke out almost at once illustrated by the brutal efforts at "ethnic cleansing" in the Balkans.

But as the Cold War ended we saw even broader examples of "tribal animosities" break out. Once the danger of an aggressive communist Soviet Union ended, rifts quickly arose among Western nations that had long been thought friends, allies, and broadly speaking, members of a common cultural, religious, and ideological tradition of free-market economics. Events leading up to the war with Iraq in March 2003 well illustrate this point. No longer facing a serious threat from the USSR that had dissolved into several separate republics, many European nations, particularly France and Germany, no longer felt dependent on the United States for security. On one issue after another they departed from U.S. leadership and struck out in directions they felt were more closely in tune with their local interests. In the case of France, an open breach took place. The French and Germans began to feel a sense of ecological release from the constraints of their dependence on the United States. Both nations now resented U.S. dominance once it had become the only global mega-power. Americans, both within the government and among the population generally, felt an acute sense of betrayal. Twice, in both World War I and World War II, the United States had come to France's rescue and "This is the thanks we get?" Trips to France and Canada sharply declined. Many Americans boycotted French wine and other products. Books were written in France that revealed virtual hatred toward all things American. One French author, Thierry Meyssan, wrote a best-selling book (in France) titled *September 11, 2001: The Big Lie.*[4] Meyssan claimed the crash of an airplane into the Pentagon on that date did not actually happen. It was all staged by the Pentagon, Meyssan claims, with the help of Hollywood. The television scenes and eye witness accounts of that event were merely proof, according to the author, of just how much trouble the Pentagon was willing to go to deceive Europe. This "lie" was simply part of the Pentagon's grand plan to take over the world. This argument takes conspiracy theory into a new and uncharted territory, but such thinking is not uncommon when one side develops hatred for the other. The first casualty in almost any cause or ideological dispute is truth. Its first cousin, critical thinking, is also suspended and no more so than among the very academics (right, left, or center) who supposedly inculcate students with the ability to think critically. More precisely, we apply critical thinking to the positions held by our opponents but likely suspend it when defending our own.

In the late Stone Age, serious disputes within villages were not long tolerated. Yet the crash of a reasonably harmonious social order among egalitarian hunters and gatherers was caused by an expanding population. Egalitarianism did continue into the very early days of agriculture. But as agriculture broke through the earlier nomadic constraints on human population, rapid growth began. Anarchy then broke out and the egalitarian social order crashed.

The evidence suggests this crash took place wherever agriculture did and forced humanity to adopt despotism as a new social ecosystem. Decision by group consensus gradually, but remorselessly, faded over a two-thousand year period. As the population continued to expand, despotism evolved from the benign to the brutal. Slavery emerged everywhere civilization did, and human sacrifice became common. When I understood this progression, in my mind Jared Diamond's argument clearly trumped that of Karl Wittfogel. Despotism had to precede the construction of large irrigation networks of a hydraulic society. Long before those networks expanded beyond local village needs, despotism of sorts had to emerge to compel construction of almost any kind of public works.

Jacques Barzun, in his book, *From Dawn to Decadence,* quotes an old Spanish proverb to this effect: "Better a thousand years of tyranny than one day of anarchy." We now turn to this process of adaptation to political tyranny in preference to local anarchy as a part of the broader FROCA process of human cultural evolution.

8

ADAPTATION THROUGH DESPOTIC CIVILIZATION

Large populations without government inevitably collapse into anarchy that small group cohesiveness usually avoids with no formal government. In large groups, as we saw in the previous chapter, violence and murder soon follow anarchy. The English philosopher, Thomas Hobbes, correctly deduced this back in the seventeenth century. Hobbes described early human communities as jungles, "red in tooth and claw," a society of "all against all." That was not usually true for hunter–gatherers as we have seen. After our shift to agriculture, however, anarchy and violence became common. The English philosopher Thomas Hobbes was constructively correct to claim that people originally created government and turned over authority to a despot as a defense against being killed by their own neighbors.[1] Again, however, that shift was usually gradual rather than abrupt and took place over generations depending on the rate of population growth.

Jared Diamond clearly describes, step by step, how despotism evolved in early agricultural society.[2] His essential outline is summarized in Figure 1. The first stage of the adaptation to a post-egalitarian social order was to select a tribal headman, sometimes called a big-man, but he had no formal political power. As the population continued to increase, a tribal chief emerged who had limited political power. Next, a political unit called a chieftain (or chiefdom) was created. Here, the chief had active political power, and in the larger chiefdoms, the office of chief became hereditary. The chief often had subordinates to help him. That stage arrived when the population reached a few thousand. The final stage to emerge on the road to adaptive despotism was the nation headed by a king, usually a god-king. Nations usually emerged when the population rose to about fifty thousand people.

Types of Societies

	Band	**Tribe**	**Chiefdom**	**State**
Membership				
Number of people	dozens	hundreds	thousands	over 50,000
Settlement pattern	nomadic	fixed: 1 village	fixed: 1 or more villages	fixed: many villages
Basic relationships	kin	kin-based clans	class and residence	class and residence
Ethnicities and languages	1	1	1	1 or more
Government				
Decision making, leadership	"egalitarian"	"egalitarian" big man	centralized, hereditary	centralized
Bureaucracy	none	none	none, or 1 or 2 levels	many levels
Monopoly of force and information	no	no	yes	yes
Conflict resolution	informal	informal	centralized	laws, judges
Hierarchy of settlement	no	no	no->*para-mount village	capital

There were of course many variations on each of these stages. A chieftain could vary considerably in size, for example. Diamond personally observed most of these stages from "egalitarianism to kleptocracy" as he terms it, on the island of New

Religion				
Justifies kleptocracy?	no	no	yes	yes→no
Economy				
Food production	no	no→yes	yes→intensive	intensive
Division of labor	no	no	no→yes	yes
Exchanges	reciprocal	reciprocal	redistributive ("tribute")	redistributive ("taxes")
Control of land	band	clan	chief	various
Society				
Stratified	no	no	yes, by kin	yes, not by kin
Slavery	no	no	small-scale	large-scale
Luxury goods for elite	no	no	yes	yes
Public architecture	no	no	no→ yes	yes
Indigenous literacy	no	no	no	often

**horizontal arrow (→) indicates that the attribute varies between less and more complex societies of that type*

from Guns, Germs, and Steel, *Jared Diamond, reprinted with permission*

Guinea, the last large primitive area on earth to come under government control. Many of its tribes were not known to outsiders until the 1930s and a few not until after World War II. The island contained all evolutionary stages from hunter–gatherer to farming and tribal chiefs, some of which were in transition as Diamond made his observations.

The more productive the agriculture and the greater the population, the more rapidly despotism evolved through these stages. As population grew so did the size of the food surpluses to be managed. Of course, there was not a leap from total anarchy to total despotism but rather an evolution via the staged process. When the population grew to the point where the informal resolution of conflicts in once close knit communities began to break down, the first step was to elect a headman who usually was a man everyone trusted, often the shaman, and his job was to help resolve disputes. He had to grow or gather his own food, had no special privileges, wore no special costume, and had to rely on trust and persuasion instead of formal authority. Group decisions were made by consensus in community gatherings where everyone could speak up. The headman's views might have carried more weight than most, people might defer to this view, but they did not have to do so. Mutual obligation still governed the distribution of resources, and no one acquired personal wealth. But when the population rose above about five hundred people, the headman system came under heavy stress and began to break down. Diamond puts the case very well:

> One reason is the problem of conflict between unrelated strangers. That problem grows astronomically as the number of people making the society increases. Relationships within a band of 20 people involve only 190 two person interactions (20 people times 19 divided by 2) but a band of 2,000 would have 1,999,000 dyads. Each dyad (a two person interaction) represents a potential time bomb that could explode

> in a murderous argument. Each murder in band or tribal based societies usually leads to an attempted revenge killing, starting one more unending cycle of murder and counter-murder that destabilizes society.... Hence a society that continues to leave conflict resolution to all its members is guaranteed to blow up.[3]

A chief of a small tribe acquired some decision-making power but was still far from being a despot. His power was limited, but a chief usually decided the outcome of a dispute. Moreover, the chief no longer toiled his own plot of ground and he was expected to spend full time on tribal business. In many cases, all the community produce was handed over to the chief who then redistributed it and took his own share in this process. At this stage, everyone was fed and no big differences in consumption were acceptable. If a big surplus occurred, a community feast was very likely and everyone shared in it. Egalitarianism was waning, to be sure, but it was by no means dead. The chief had to use his authority with discretion and would not make controversial decisions by himself. Persuasion was important and a council of elders usually made the key decisions, and they usually reflected a broad consensus. As small tribes grew larger however, chiefs acquired a monopoly on the power to resolve disputes. Meanwhile, a growing population brought forth a new need, namely a requirement for public works. These included large meeting halls, places of worship (and the chief was often shaman or had been one) and even such things as small bridges over creeks or streams. These works were "paid" for out of the community surplus, but in some cases tribal members were called upon to provide work, not product. Meanwhile, if conflicts with neighboring clans or tribes took place, the chief often took direct charge. Raids on neighboring tribes yielded prisoners who became the nucleus of a slave class. As the population of a tribe expanded, the need for slaves rose. They would increasingly be put to work doing jobs others didn't like, replaced the labor drafts to do public works, or served as personal servants of the chief and his family members. Slaves were often treated reasonably well, unless they misbehaved or tried to escape and made up only a minority of the population at this time.

Once a tribe had grown too large for a simple chief with limited power to administer, the stage was set for the evolution of a new political entity, called the chiefdom, usually a collection of villages and with perhaps a thousand or more people. In a chiefdom, the chief acquired considerably more political power, and his office usually became a hereditary right limited to his family. The logic of succession through heredity was that it reduced conflict, indeed open warfare, over who became the next chief. Still that step gave rise to an elite class, the nobility, and it began to evolve again. Nobles became made subchiefs at first, an office created to watch over smaller sections of the tribal territory. Social stratification thus emerged if only at first into commoners and nobility. Slaves became more numerous and formed an official underclass with few of the rights of tribal members.

As population continued to grow, the division of labor became even more narrow. Specialized occupations evolved, because when productive surpluses became substantial, the number of these specialties expanded, in part to cater to the more sophisticated demands of the new elite. The chief and his staff took in all that was produced and redistributed it to the people. This redistribution included items that were crafted rather than grown. The chief's village then began to acquire the attributes of a capital city. In the larger chiefdoms, a clear distinction began to emerge between the living standards and dress of the elite at the top and of commoners and slaves at the bottom. More and more of the community output remained in the hands of the elite and their families. Their greater share enabled them to live on a higher standard and yet they rarely participated in any kind of production. Additionally, the commoners might be required to engage in "respect" ritual behavior toward the elite, such as bowing or using deferential language not unlike subordinate animals show to the dominant members.

This severe departure from the egalitarian traditions from

which chieftains emerged demanded justification. Religion usually served that purpose. First, even in a small chiefdom, a tribal genesis myth and perhaps some kind of ideology or theology to go with it had long provided life with a spiritual dimension. It often involved voluntary submission by people to the one or more higher powers they had long chosen to worship. This was the start of institutionalized religion. Chiefs discovered that preexisting tribal spiritual beliefs could buttress their authority since they were often built on the earlier role of the shaman. As mentioned earlier, the chief might have been the shaman and so the transition of shaman–chief into the earthly spokesman for those higher-powers-that-be was often relatively smooth and seamless. Once he began to intervene on behalf of the broader community to bring rain or cure the sick, he could logically claim special power and privilege as well. Thus, a chief in a larger chiefdom deserved, and often received without much resentment, a show of deference. It had long been given freely to the Higher Power and so the chief, as the representative of that Higher Power here on earth, deserved to be shown similar respect behavior. The "divine right of kings" was accepted logic.

When a population reached about fifty-thousand people, the chieftain evolved into a new unit, the nation. The top man (and occasionally woman) was no longer merely a chief who had the ability to intervene with higher powers. As the head of a nation that combined several (often related) tribes, the top man was the chief-of-chiefs, the king, and often as not, the god-king. He had become a fully divine personage in his own right. As such he might remain aloof from the masses and was now the absolute ruler, a despot in every sense of the word. In many cases communal ownership of property that was traditional in tribes or clans from the earliest days of agriculture now became the sole possession of the divine presence who graciously allowed his minions to farm the land provided they turned over most of it to the god-king. (A unique and interesting variation on this theme emerged among the Incas where the title to land was held by the dead.)

The king or god-king had a complete monopoly on power, whether in the realm of economic activity, religion, politics, military and foreign affairs, or police power. Hereafter we will refer to him simply as the despot. Above all, the aim of this fully fledged despot was to maintain his control and preserve stability in his realm. Any challenge to that control or threat to stability was typically dealt with ruthlessly. So, with the arrival of the despot in some region, came the equilibrium. This stasis followed more than three thousand years of punctuation, of new frontiers opened by agricultural technology, the rapid growth of population and surplus wealth, which in turn flowered into a wide range of civilization-enabling tools, weapons, and means of transportation—each presenting new exploitable opportunities.

But with this flowering came the constant threat of anarchy and social turmoil leading to violence and murder. The evolving cultural ecosystem had to struggle with this challenge that had caused the egalitarian society to crumble or crash. The effort to maintain order in particular communities is reflected in the evolution of the headman into a supreme despot in control of the local cultural or social ecosystem. Safety was much better, but much personal freedom vanished. So did any free wheeling technological opportunity wherever a despot reigned. Despots emerged to preserve order more or less with the implicit consent of their subjects. They may well have become cruel and self-indulgent kleptocrats who kept their subjects in poverty, but they did preserve order.

Let's note, however, the sharp departure from self-organization, the primary way that nature balances an ecosystem. No forest, prairie, or meadow has a "boss" who specifies and controls the behavior of others. Neither did humans have that in their hunter–gatherer culture with its egalitarian social order. But with the arrival of the technology that enabled humans to produce their food and with large surpluses, things changed. Surplus food and settled village life fostered the rapid growth of human population that, in the absence of government and political authority, led to anarchy, chaos, and violence. That proved so traumatic that people to created authorities to preserve order and harmony. Large populations led to the creation of supreme despots who had the authority to apply coercive force to maintain law and order. They largely succeeded. But with that success human culture emerged into a new territory. For the first time, evolution had ended the punctuation stage of rapid change by creating despotism rather than purely competitive self-organization to achieve stasis with its order and equilibrium. One policy hallmark of stable social ecosystems is *"a place for everyone, and everyone in their place."* That place may be humble, and typically was for 90 percent or more of the population. Still, that policy clearly helped to maintain social stability. It discouraged ambition and innovation and usually came with an ethic that stressed duty and obligation far above personal freedom. Charles Dickens makes this point: "Let us bless our Lord and his relations, grateful for our daily rations, content within our proper stations."

Religion plays a very prominent role in such acceptance and often comes with minutely specified prescriptions for the details of everyday life. Ritual politeness plays a major role, and strict rules of deference apply to those below on how they should behave in the presence of those above. To go above one's station was often seen as a punishable offense. Dress codes often applied and were strictly enforced. Such obligations were (and are where they still apply) justified to thwart the danger of anarchy and the mayhem that comes with it. Moreover, in any highly stratified society with a small and highly privileged elite, that elite was (and is) ever mindful that even minor lapses of deference behavior and acceptance of place can cascade toward anarchy (or we might say, ecological release by the masses) with the slaughter of that elite soon after. A policy that did survive from the egalitarian era, however, was the notion of dividing the economic pie into traditional shares. No one was expected to compete with other people to find a place and earn a living. One more or less knew what to expect as one was growing up.

The young simply entered into the same occupation as their fathers. For the great bulk of people, that occupation was farming. In cities, one's occupation was often artisan or merchant. For slaves' children it was slavery. In most ancient civilizations some form of the caste system arose—social stratification based on the ascribed relative social worth of one's occupation or function. Priests usually worked out the rank order and placed themselves at or near the top. This rank order was then presented as either "natural" or "god-given," and in general even the people at the bottom believed it, often including slaves. There were many variations on this theme. Plato's division of people into the gold elite, the silver sub-elite, and the bronze masses who actually did real work was widely accepted as a natural state of humanity.[4]

Writing was the major innovation of early civilization. The first example we know about was the cuneiform writing of Babylon. Egyptian hieroglyphics and Chinese script followed later. Starting around five thousand years ago, despots needed writing as a method of keeping track of how much grain was produced, how much was collected as tribute or taxes and for other purposes, which is why numbers came before words. It was a vital control device aside from being a form of communication. Knowledge of writing was closely held long after it was invented; only the elite was privy to its mysteries for nearly two thousand years after its invention. That control did not begin to fade until the invention of the Greek alphabet about 700 BCE. At first, writing and its script functioned almost as code for the official business of the state. Moreover, being literate in the awkward symbols of early writing was clearly a skill that helped justify one's position as belonging to the elite.

Writing, together with bronze metallurgy, were the two last major innovations at the dawn of civilization. After that, not much new technology surfaced for several thousand years. The two major exceptions were the Greek alphabet in 700 BCE and iron metallurgy around the same time.

9

EQUILIBRIUM THROUGH REPRESSED TECHNOLOGY

As a college student taking world history, I was always puzzled about why neither the Romans nor the Chinese had invented such things as the printing press, steam engine, or cannons. None of those things seemed to challenge their basic understanding or practical know-how. If they could build great walls as did the Chinese or great roads and aqueducts as did the Romans, surely they could have invented those reasonably straightforward technologies. But they did not. In fact, as I mentioned in the last chapter, and with the exceptions mentioned, not much really new was invented from the dawn of civilization to the late feudal era, a period of nearly five thousand years. From the standpoint of technology, this was truly a period of stasis and equilibrium. Today we think of new technology as a self-reinforcing or self-accelerating process, or as some put it, autocatalysis. Many people now assume technology cannot be stopped.

Yet, for the five thousand years when ancient civilizations dominated by despotic social orders ruled, technology all but stopped advancing. Civilization put the brakes on new innovation because innovation had become problematic. Improvements to existing technology were sometimes made, of course, but very little new emerged. An ox-cart from 3,000 BCE was not much different from one made in 1800 CE.

In fact, average material standards of living did not rise at all for those five millennia. Descriptions of life in the Bible were descriptions modern people could relate to until quite recently.

Indeed, for those living in more remote rural areas, the Biblical lifestyle persisted into the early twentieth century. Economic historians such as Cameron put the issue this way:

> In terms of technological development the record is extremely sparse. Almost all the major elements of technology that served ancient civilizations—domesticated plants and animals, textiles, pottery, metallurgy, monumental architecture, the wheel, sailing ships, and so on—*had been achieved before the dawn of recorded history.*" (Emphasis added)[1]

Although there were some notable examples of outright suppression of technology, other factors played a more important role. The first of these was slavery. Closely related to slavery was a system of social stratification based on any of several variations of the caste that systematically kept the status of most forms of physical labor very low. In the households of the elite, servants, who were often as not slaves, did all the household work. The master or mistress of the household would demean themselves if they were to cook, wash, clean, repair damage, or even garden (possibly excepting flowers).[2] I recall a time in the early seventies in Jakarta when I agreed to house-sit for an Australian diplomat and his family while they went on home leave for a month. Their very large house had a staff of seven, including a chauffer, cook, kitchen helper, two maids, a houseboy, and a gardener/watchman. Being single at the time I was used to having a maid of all work from my days in Bangkok, but at least I felt free to do things for myself when I felt like it. But in that house in Jakarta, I could do nothing. I could not so much as brew myself a cup a tea or draw my bath without demeaning myself and questioning the worth of the staff at the same time. Feeling like a prisoner of the local status system, I was much relieved when the family returned. William Manchester in his biography of Winston Churchill noted that Britain's greatest statesman probably never once drew a bath for himself or had to dress himself except perhaps as a young officer in India.

Today in most of Western culture—with the exception of the very wealthy—being waited on hand and foot is rare and many people take pride in being reasonably self-sufficient in the daily routines of life. Yet, for anyone of means self-sufficiency is a very new ethic. Into the early years of the twentieth century, household workers—or domestics as they were called—comprised the largest category of hired employees in America, Britain, and most of Europe. That would soon end for two related reasons. New and better industrial technology opened up new and much better jobs in industrial corporations on one hand, while a surge of new and better domestic appliances such as vacuum cleaners, electric dish and clothes washers and dryers, toasters, irons, and so on took much of the drudgery out of housework. At the same time, the telephone and typewriter drained domestic workers from households into business firms and government offices at a frightful rate after about 1910. In 1910 most secretaries were male, and only ten years later they were mostly female. By the end of World War II all but the very wealthy had to learn a reasonable degree of self-sufficiency in the home, aided of course by all that new technology.

Again, much of this change took place in the span of one lifetime. Until early in the twentieth century little menial work, particularly around the house, was done by people of means. This ethic was a lasting legacy of the social stratification imposed by all ancient despotisms. Servitude in various forms, however, did systematically transfer wealth from the upper to the lower ranks of society. Slaves had to be fed and cared for to at least some minimal extent in a domestic household as did servants who might also earn money. Household slaves and other servants did not produce wealth so much as help the elite consume the wealth produced by others.

At the upper strata, however, lived people who worked with their minds in one way or another, such as priests, scribes, government officials, and philosophers. They put their high status at risk, however, if they performed "demeaning" physical work beneath their dignity. Because most inventive thoughts about new tools or artifacts will go nowhere unless someone transforms ideas into useable hardware, such transformation means physical work. If one avoids physical work, creative ideas about how to do it better would not likely come to mind. Slaves did most of the labor that might lead someone to think of a better way of doing a job, but slaves had no status. If they did invent something, they almost certainly would not benefit from it, even if it were adopted.

First, why would a slave owner even care about labor-saving improvements? In fact they did not. Economic historians such as Rondo Cameron have long noted this fact. Owning slaves was a sure sign of high status so why invent ways to reduce their number and one's status at the same time? Why bother to make it easier to do work that slave owners themselves never had to do? Moreover, slave owners might well incur the wrath of their fellows by promoting labor-saving devices except under very unusual circumstances.

Ancient civilizations such as Rome did design and build many kinds of public works that conserved labor such as paved roads and aqueducts. But the Roman aqueducts did not displace masses of slaves who had previously been forced to carry water in buckets over very long distances, which was never done (because the slave would probably have had to consume most of the water he carried to avoid death by dehydration). Instead, slaves (and soldiers in the case of Rome) were put to work building roads and aqueducts that allowed for larger cities with even more slaves.

An exception to the ban on physical effort by the elite were the physical skills of warriors and athletes. Strong healthy bodies were much admired because physical strength was needed to excel in both activities. The Greeks held both sets of skills in high regard as evidenced by the Olympic Games that included warrior events of archery and javelin throwing.

Another partial exception to the strong disincentives to innovate came when a rival tribe or other power had a tool or weapon that could be used to defeat the local tribe. The local power would then adopt that innovation as a matter of self-defense and might at times improve upon it. This defensive response was, in fact, the main avenue for the spread of such innovation as took hold. The defensive adoption of external innovations, however, did not undermine the ancient status system based on social stratification and the caste system.

Sometimes an innovation adopted to counter a competitive threat or gain a military advantage might later be reversed if it did threaten the status system or the stability of the social order. A good example comes from Japan. The Japanese social order was stratified into a caste system with the samurai warrior aristocracy on top. Farmers came next, followed by artisans and merchants. At the very bottom were the etas, outcasts who were the untouchables and who performed all the labor such as butchering that others considered unclean. The etas were not technically counted as slaves, but they served a similar social function. They took certain kinds of undesirable and ritually unclean work off the backs of the higher castes. In sixteenth-century feudal Japan, the major samurai clans such as Satsuma were constantly at war with each other. Then in 1543, two Portuguese missionaries arrived in Japan bringing with them two arquebuses, an early version of the musket.[3] The utility of these firearms were at once evident to Oda Nobunaga, a clan leader who wanted to unify Japan. Under his sponsorship, the Japanese "reverse engineered" these two guns and began making an improved version of their own. Nabunaga also began training his soldiers in their use. He died in 1582 and shortly thereafter Ieyasu Tokugawa took up the effort to unify Japan. By 1600, moreover, Japan had more and better muskets than any other nation on earth.[4] Then, in that year, Tokugawa won the battle of Sekigahara that put an end to these constant civil wars. Tokugawa Ieyasu became shogun in 1603 and established his long-lasting Tokugawa Shogunate that endured until 1868.

Although Tokugawa had used the musket to win his victory, he was nevertheless disturbed by it.

In fifteenth-century Japan, the sword served as the main symbol of samurai dominance, and the ability to use it well was highly prized. Yet, if any farmer, or even a lowly eta, with musket in hand could gun down a samurai warrior from a safe distance, such a weapon posed a serious threat to the social stability of the entire caste system. Fearing this threat, the shogun began collecting all the muskets, had them destroyed, and later shut down all the workshops that had made muskets. By 1640, few muskets remained in Japan and the samurai class remained firmly at the top, its prestige intact.[5] Why could the shogun get away with that? Why could not rivals make their own muskets? First, Tokugawa forced his potential rivals—clan lords and their families—to reside in his capital where they lived as potential hostages. Thus the shogun took firm control as the top political authority in all Japan and effectively had no internal rivals. He could easily quash any move rivals made to acquire muskets to challenge his authority. To assure no challenges came from outside Japan, the new shogun also banished all outsiders and made it a serious offense for Japanese subjects to travel abroad. Most Japanese who did leave Japan would fear to return and hence posed no threat. That prohibition stayed in effect until after the arrival in Tokyo Bay of the American Commodore Perry in 1853, more than two hundred years later.

A similar example of outright suppression of new technology comes from China in the fifteenth century. For years the Chinese allowed foreign merchants and their ships to handle foreign trade, but in the Ming dynasty Chinese themselves began to trade and build their own ships. They traded at first with the Philippines, Japan, and the Malay Peninsula, and between 1415 and 1433, a Chinese admiral, Zheng He (Cheng-Ho), had acquired a large fleet of ships and finally assembled them in a huge convoy that carried 28,000 people all the way to the east coast of Africa. However, as Rondo Cameron states:

> Then suddenly in 1433, the emperor forbad further voyages, decreed the destruction of ocean-going ships and prohibited his subjects from traveling abroad. The [Chinese] colonies were left to wither away. One wonders how the course of world history might have differed had the Chinese still been in the Indian Ocean when the Portuguese arrived at the end of that (fifteenth) century.[6]

The emperor was the sole authority in China and so he could impose his will. To make sure he kept control, he also forbade his subjects to travel abroad. That prohibition also aimed to preclude the "middle kingdom" from exposure to foreign barbaric and corrupt influences. At that time, China was arguably the most technologically advanced nation on earth. With some justification at the time the Chinese felt they had nothing much to learn from the foreign barbarians. That time, however, was about to pass.[7]

We will see in Part III why European despots of the fifteenth century era could not safely suppress the new surge of technology that was then starting to take place in Europe. That technology was indeed proving highly disruptive to the feudal social order but competition between the despots made it impossible for them to match the policies of the Tokugawa in the seventeenth century or of the Chinese Emperor in the fifteenth century.

Even as despotism arose to suppress local anarchy, many despots soon felt forced to become imperial to maintain order. They were faced with almost constant border warfare with other states or chiefdoms that were often rivals for the same scarce resources. Thus, nations evolved into despotic empires as a part of an extended phase of adaptation to achieve stasis and stability in human ecosystems. But this brought a new problem of how to control diverse tribes with different religions and cultures.

10

EMPIRES EVOLVE TO QUELL INTER-TRIBAL WARFARE

The evolution of despotism kept murderous anarchy partially at bay within the growing population of multitribal nations or chiefdoms. But there were many such agricultural communities growing and evolving at the same time. Disputes between rival communities erupted for the same reason they had earlier among family units in the same clan or tribe. No recognized political system existed that could maintain law and thus keep order without resorting to violence. The emerging nations had no common authority that they could use to resolve disputes and enforce decisions. When nations and chiefdoms were forced to compete for the same scarce resources, they often resolved their differences by combat, and the strongest usually prevailed.

Thus, frequent border clashes were routine. Attacks were followed by counterattacks, much like the intratribal family feuds. Where anarchy prevailed at any level, killings followed by revenge killings became standard. Anarchy scales up almost seamlessly from families to tribes and on to nations and empires. The constant clashes between Israel and Palestine over the past half century illustrate the point. Meanwhile, in the ancient world, constant border clashes added social status to tribal warriors. Higher status often led to the rise of a warrior class with a strong ethic usually based on such values as loyalty, courage, honor, and duty to tribe, king, or country as with the Spartans, samurai, or Romans. Border conflicts and human-to-human predation were common even among nomadic hunter–gatherers. Predatory behavior dates back hundreds of thousands of years. Still, the small nomadic populations were always on the move. They also had ample space they could use to keep safe distances between rival bands. Thus, the occasional skirmishes between hunting bands scarcely amounted to warfare. Moreover, nomadic Stone Age bands were far too small to sustain serious blood letting and still survive.

But when the nomadic way of life gave way to permanent agricultural settlements, based on stored wealth and with much larger populations, endemic warfare became a standard feature of human existence. War could resolve disputes among rivals when diplomacy failed. Standing armies (and navies for nations or city-states located along a coast) became more necessary for the defense of the realm.

The competition for scarce resources, however, was not the only problem for emerging nations. Constant intertribal border raids, clashes, and warfare between nations disrupted vital trade. Trade had early evolved in response to the growing division of labor that large agricultural surpluses both required and then enabled. It allowed a nation to exchange its surpluses for those things in scarce supply. As Cameron has put it, "The division of labor and the evolution of new crafts such as pottery and metallurgy, required some form of trade and commerce."[1] In fact, civilizations could hardly have arisen without trade.

Sargon of Akkad founded the first empire about 2300 BCE. It was located below present-day Baghdad in the Tigris and Euphrates valley. The valley was rich and fertile because of the constant renewal of river silt, but it lacked many other resources needed in the cities of ancient civilizations, such as timber, stone quarries, and ore for smelting. Thus the cities located in such valleys relied on trade to supply its other needs unless it was strong enough to seize and defend a nearby source of supply.

Other empires with similar needs for trade followed Sargon's. Babylonia replaced Sargon's empire and became much larger. The Hittites created an early empire of their own. So did the Egyptians, Assyrians, Persians, and Greeks, most notably the city-state of Athens. Then the Macedonians led by Alexander the Great quickly took over all of Greece, Persia, and Egypt circa 330 BCE. But, the Roman Empire would rise beyond them all. Rome began as a small city-state about 400 BCE and by about 200 BCE began to expand after defeating its rival, Carthage. By circa 200 CE Rome controlled the entire coastline of the Mediterranean Sea, including its inlets. Rome also controlled most of the Balkan Peninsula, all of Iberia (modern Spain and Portugal), France, England, and Turkey, as well as all of present-day Iraq and Armenia and later most of the island of Great Britain. Rome remained the dominant power in the area from circa 150 BCE to 400 CE but collapsed in 476 CE after a little more than one hundred years of decline. Rome reached its apex of power about 200 CE.

The eastern half of Rome's later empire, based on Constantinople, survived Rome's collapse but finally fell to the Ottoman Turks. They captured the city in 1453. Elsewhere in the ancient world about the same time, China under the Han emerged to become Asia's greatest empire. Other empires arose in India, such as the Gupta Empire encompassing the northern states. Later in sub-Sahara Africa came the Bantu to dominate the Kush and Axumites. First the Maya and later the Incas and Aztecs created empires in the Americas.

Thus, as the nations and city-states became larger and more numerous, the anarchy between them became evermore problematic. The desire to end this anarchy became a major motive in the creation of all ancient empires. Empires evolved as the next stage of a process of adaptation that began with the emergence of the first headmen and then chiefs. Trade itself, however, had become a "new frontier" of its own. Opportunities for predation by outlying hill tribes or chiefdoms forced to farm or herd on marginal land expanded as nations and empires arose. Trade caravans or ships presented good opportunities for making a better living through brigandage and piracy for tribes that lived along trade routes. They could raid caravans in the desert spaces between nations or seize ships at sea and then vanish. A nation, if it wanted to survive, had to protect its trade routes, but that was expensive.

The Phoenicians of the eastern Mediterranean Sea created the first maritime trading empire (of sorts) primarily by founding colonies along the African coast, as well as a few in Spain and in Sardinia and Sicily. One of these colonies, Carthage, located in present-day Tunisia, later established an empire of its own to compete with Rome. But after some initial success by Carthage, Rome crushed Carthage and "sowed the land with salt." The Phoenicians organized themselves mainly as autonomous city-states. Sidon and Tyre were the most famous. These cities survived first by paying tribute to the more powerful nations and second through their extensive trading connections. The Phoenicians provided a service other nations needed. Interior states or empires need not create their own navies and merchant fleets if they could rely on the Phoenicians to bring cargo to them.

The Athenians and other Greek city-states also colonized extensively, mostly in Europe along the eastern shore of the Mediterranean and all along the Black Sea coast.[2] Trade thus played a major role in early Greek success and the Athenian Navy played a vital role in defeating a much larger Persian empire. From an ecological point of view, as trade expanded in the ancient world, nations and emerging empires facilitated the physical recycling system of expanding human ecosystems in ancient times. However, these ecosystems could and did crash as we shall see.

Still, physical goods were not the only things exchanged in the course of trade. So too were ideas, ideologies, and religions along with occasional technology. Traders from many regions brought goods to central collection centers, such as the Greek city of Melitus, intermingled and exchanged ideas. The need to keep accurate records, moreover, gave rise to the need for a better form of writing and led to the invention of the Greek alphabet. The alphabet promoted broad literacy and for the first time writing lost its role as a code known only by a specialized elite. Literature, philosophy, and science all emerged in Greece partly as a result of the alphabet.

Further, as traders and Greek scribes compared the various tribal creation stories the realization arose that they could not all be true. That insight led Greeks such as Thales, Zeno, and Hericletus to speculate on the nature of the world through reason alone. They began to reject explanations handed down through tribal oral history, even those of Greece. The alphabet indeed became the origin of the first serious challenge to despotism, but one that would take many more centuries to take real effect.

The alphabet was also the beginning of a division that continues today between the insights of Athens and those of Jerusalem, that is to say, between the reason of Athens and the faith of Jerusalem. But again, the major impact of this division did not come for another two thousand years as we shall see.

The cascading consequences of those early exchanges between merchants could be destabilizing. Almost every tribe at some point had adopted its own tribal creation history that featured its tribe at the center of all creation. It thus became hard—especially for a god-king who presumed to be the sole earthly representative of the higher-powers-that-be—to maintain the logic of his monopoly on and divine claim to despotic power. If adjacent states held alternative beliefs, those states stood as potential threats. Thus, through the advancement of intertribal communication, a god-king acquired a motive to dominate other states and impose his religion on them before his rivals conquered and imposed their religion on him. So almost as soon as nations emerged in the ancient world, the larger nations morphed into empires by taking over adjacent rivals. Such takeovers were also motivated by the quest to create stable social orders safe from external threat or from the internal disruptions brought about by newer technologies as the alphabet and papyrus for writing.

One new approach to achieving this stability was the rise of universalized religions. A single religion could help otherwise diverse cultures to share a common empire. Western monotheism, One God, supreme over all, began to replace local tribal religions beginning with the Hebrews who united their twelve tribes under a single God. To quote Karen Armstrong, "a single deity who was the focus of all worship would integrate society as well as the individual." This insight, she points out, would later drive Muhammad as well.[3]

Rome tolerated a variety of local religions provided they acknowledged Rome's overall political supremacy. The problem monotheism posed for Rome was that its logic did not permit its adherents, mainly Hebrews at the outset, to recognize any other gods but God. Although Judaism was favored early in the Roman Empire, the Hebrews continued to resist Roman rule. They longed for the day when a Messiah would restore an independent kingdom such as David's. In the end, Rome displayed its ruthlessness by oppressing Judaism hard, destroying the Temple in Jerusalem in 77 CE, and forcing the Jews to disperse.

Just before the Jewish diaspora from Rome, a new universal and monotheistic religion arose out of Judaism, namely Christianity. Like the Jews, the growing Christian religion rejected any other gods but God. Thus, the Romans began to persecute the Christians as well. Even so, the new faith grew rapidly and began to spread throughout the empire. By about 330 CE, the Roman Emperor Constantine became a convert after winning a battle for control of Rome and made Christianity the official religion of Rome. The Christian faith prevailed, however, only after Rome had endured considerable stress.

The Roman Empire came closer than any other to achieving the long sought goal of a universal law and order. By about 250 CE, Rome was huge. It had managed to maintain law and order within its realm by crushing any who might challenge its power, as did the Celts, Greeks, or Hebrews of Palestine. By 200 CE, a person could safely travel from one end of Rome's Empire to the other without serious threat of piracy at sea or of robbery along Rome's extensive networks of paved roads. These roads remained the best in the world for nearly fifteen hundred years after Rome's collapse.

But collapse Rome did, because it became less internally cohesive as its empire continued to expand. Moreover, Rome never became strong enough, large enough, internally cohesive enough, or technically sophisticated enough to subdue or dominate all the border tribes. These tribes continued to raid, to probe, plunder and pillage, and in the end these clashes depleted the Roman resilience until the empire finally collapsed. That collapse was the crash of the largest human ecosystem yet formed. As had previous disturbances, that crash would later open new frontiers. It would also create new opportunities to innovate that led to an explosion of new technology that had been long suppressed. The collapse of Rome thus would spell the beginning of the end to ancient despotisms everywhere. Indeed, Rome's collapse would launch a new FROCA process that continues to this day.

So, to that crucial turning point in cultural evolution, we turn next in Chapter 11.

11

BARBARIC DISRUPTION: THE FALL OF ROME

A species that migrates to a new ecosystem can prove highly disruptive. The migrant species is calibrated in terms of fitness to its native ecosystem, not the new one. It may thus do great damage to its new home because it lacks the necessary constraints. When the first human immigrants came to the islands of the Pacific, from New Zealand to Hawaii and all in between, their presence disrupted the ecosystems such as Easter Island as discussed earlier.

The tribes living along the extensive borders of the Roman Empire provoked a similar kind of disruption to Rome's cultural ecosystem. The invaders destroyed Rome's central control and were thus unconstrained by it. By, stages these invaders brought down the aging and debilitated Roman Empire in the west. The Greco–Roman civilization had long referred to Europe's northern tribes as barbarians, a term that initially meant "foreigners" but came to mean "less civilized." Because they were the first, the Germanic tribes proved to be the most disruptive.

For years, Rome's powerful legions had served as the empire's "immune system," and kept incursions by barbarians at bay. Roman legions were feared everywhere they marched. In England, however, Hadrian's Wall in the north had halted invaders from Scotland. Rome had expanded over the years by invading various tribal regions, taking them over, pacifying them and sometimes Romanizing them. It had done this successively in Iberia (Spain), Gaul (France), England, in the Balkans, and elsewhere. Nations that resisted and were defeated in battle became a rich source of slaves and sometimes raw materials and gold. But where the locals remained peaceful and compliant, they soon integrated into Rome's system of law and order and might eventually become Roman citizens. Compliant regions often became prosperous, willing parts of the expanding Roman Empire, and a growing source of recruits for the legions.

But by 200 CE, Rome's empire had become so large and extensive it was forced to recruit provincials and tribesmen along the borders. These recruits often made good soldiers, but in the end such cultural diffusion diluted the loyalty and effectiveness of Rome's legions. Those soldiers became less an elite army of faithful citizens enthusiastically defending king and country from the barbarians to occasionally fighting on behalf of their own tribes. As an army of paid mercenaries, some soldiers held tenuous loyalty to Rome and owed their primary allegiance to their native tribes that Rome was fighting.

Again, the most dangerous of these tribes by far were the Germans. Rome had been unsuccessful in penetrating the thick forests of Germany and so more or less stopped at the Rhine River. The Germans were fierce warriors, and Rome would occasionally recruit soldiers from German tribes. This policy was to prove fatal. In 9 CE, one recruit, Arminius, rose to high command. He understood Rome's tactics and battle strategy, but he became angry at Augustus and defected back to Germany. About this time German tribes had caused Rome enough trouble that Rome's local commander, Varus, planned to invade Germany with three full legions in retaliation. Arminius knew of these plans and that the Roman legions were most vulnerable when shifting camps and marching from one sector to another. Thus he planned his attack to strike during such a shift and in a dense part of the Tuetoburg forest where the legions could not form into fighting formations. It worked brilliantly. All three legions were wiped out, the worst defeat Rome's legions had ever suffered, and it marked a turning point. Rome would expand no more into Germany.[1] Rome continued to expand in the east, and its empire remained powerful for another three hundred years. Still, a high point of sorts had been reached. By 300 CE, barbarians everywhere increasingly pressured a weakened Rome, which began to evacuate secure regions such as England. Rome could no longer afford the high cost of their defense. The quality of new recruits also declined as the legions became more mercenary. Emperors raised taxes to extortionate levels to pay first for the kleptocratic lifestyle of the social elite and second for defense. Luxury was the last thing to go, and many legionnaires were diverted to collecting taxes.

Toward the end, Emperor Diocletian tried to reform the bureaucracy but it soon became onerous and rule-bound. The multi-layered system encouraged people to pay bribes to bypass Roman law and escape taxes. Thus, corruption became pronounced and the institutional efficiency of Rome's rule faltered, and it continued to lose control of border areas. Major raids and tribal incursions into Roman territory became more common. Rome was now forced to pay bribes and tribute to barbarian chieftains—such as Attila the Hun—to hold them off. The cost of such tribute in turn forced taxes to rise higher yet and so Rome's downward spiral continued.

Earlier, Rome's victories over the border tribes had produced a steady flow of new slaves, but that source diminished when Rome's victories ceased. Therefore, the ongoing maintenance of roads and aqueducts began to falter. And so on. In 410 CE, the Visigoths led by King Alaric drove all the way to Rome and sacked the city, although they did not devastate it. Meanwhile, the center of imperial power had shifted to Constantinople (now Istanbul).

In 476 CE the western Roman Empire finally collapsed, marking the end of ancient history. A semblance of Roman civilization continued under barbarian rule for a few generations. By about 600 CE, however, the Dark Ages descended. Literacy went into decline. By about 700 CE most of the barbarians could neither read nor write. Literacy would disappear in a generation or two, while piracy and banditry, meanwhile, rose sharply.

Travel became dangerous except in heavily armed groups or convoys. By 700 CE, all pretense of central and even provincial authority in the western empire had vanished. The population of Rome fell from a million people at its peak to about fifty thousand at its low point. With a few exceptions, such as Venice, cities all but vanished. North of Italy they did vanish. London and Paris, fairly sizeable places under Roman rule, lost most of their population as the Saxons invaded from Germany.

Political authority became highly fragmented. Thousands of feudal manors evolved from the large slave-holding estates of the Roman elites and these manors held most of the population. Roman slavery, however, had already morphed into serfdom and feudal-type manors had begun to appear. Serfs were not slaves under feudalism but they had little freedom. They were bound to their lords of the manor, but the lords were also bound to the serfs, who acquired tenure and so could not be sold off. In effect the serfs struck a bargain of exchanging personal freedom for security. Those displaced from the cities would enter the manor to gain security but they also gave up their freedom. In any event, slavery required constant re-supply because the birth rate of slaves was so low they did not fully reproduce themselves. As Cameron pointed out, the number of slaves would tend to decline over time, and as the power of Rome declined, its sources of new slaves dried up.[2]

Yet Rome's influence did not entirely vanish. The Catholic Church survived the collapse and became the one culturally unifying institution of Western Europe, in part because Christian missionaries had penetrated into the barbarian realms even before Rome collapsed or the Dark Ages fully descended. For example, St. Patrick had helped convert Ireland, never a part of the empire. As a result, monks from Ireland roamed over Western Europe setting up monasteries and keeping the Latin language alive, since the Church's liturgy was in Latin. Meanwhile, priests took up the occupation of copying bibles and other Greco–Roman works in Latin to preserve them for the future. Monks became almost the sole repositories of literacy in Western Europe and made the Latin language its lingua franca. Monks from any part of Europe continued communicating in Latin even as tribal or "vulgar" languages took over daily conversation among most of the population.

Meanwhile, those parts of the Roman Empire in North Africa more or less continued to live in the Roman fashion even as Rome's authority fell apart. But a new, and in some ways more sinister, threat to the now Christianized Roman world erupted. Soon after the Germans and the Vikings took control of Europe north of the Alps, nomadic Arab herdsmen of the Arabian Peninsula invaded the east. These Arabs were fierce warriors who raided border provinces on occasion. But they were also highly fragmented into often warring tribes. They had no central authority. The two largest towns were Mecca and Medina, located near the Red Sea coast a few hundred miles below the Sinai.

In 610 CE a new monotheistic religion called Islam suddenly began to flower in Mecca. Muhammad was its prophet, and the holy Koran was the scriptural message God revealed to him through God's messenger, the angel Gabriel. While Islam recognized both Jews and Christians as "people of the book," Muhammad insisted the Koran transcended all the earlier scripture. Islam, moreover, had a distinctive Arab flavor. No quotations from the Koran in any language other than Arab were regarded as official, and that constraint remains in effect today. As Karen Armstrong has pointed out, *"Muhammad knew that monotheism was inimical to tribalism: a single deity who is the focus of all worship would integrate society as well as the individual."*[3]

Muhammad, initially at least, was correct. Very quickly, Muhammad unified the Arab tribes under Islam. In Muhammad's lifetime, unity apparently gave the Arabs an enormous sense of ecological release from the constraints of their previous infighting. The Arabs quickly began exporting their faith through invasion of Roman lands. Islam aimed to unify all the people of the world in that faith, much as the Christians hoped to do. The time was ripe. In North Africa, the Roman civilization was debilitated and nearly helpless, the legions long gone. While Egypt was under the authority of Constantinople rather than Rome, Egypt was not well defended. While the eastern empire was victorious in a brutal struggle with Persia, that war had left both empires very weak. Neither proved a match for the newly energized Islamic Arabs.

By 638, the Arabs had seized Jerusalem. By 641, they had conquered all of Palestine, Egypt, and Syria. They also made deep inroads into the Persian empire and later took it over. By 650, the Muslims ruled Iraq, Iran, Afghanistan, and much of present-day Pakistan. Tripoli in Libya had also fallen along with the island of Cyprus by that date. After a short pause to consolidate their gains, the Muslims overran what remained of North Africa then captured Spain between 705 and 717. In less than one hundred years Arab Muslims had created a vast empire to rival Rome's. In 732, they crossed the Pyrenees from Spain and penetrated deeply into France. They were stopped at the battle of Poitiers and turned back by the Franks under Charles Martel. In Anatolia, the Byzantines put up a better defense and thus retarded Arab expansion. But throughout Africa and much of the Middle East and North Africa, the Christian faith all but vanished, as its adherents began to convert to Islam.

Still, the Arabs never unified their vast empire as well as the Romans had done at their peak. The Arabs reached their high water mark quickly, between about 800 and 900 CE. Internal battles began early and were never completely resolved. A basic split that erupted early on remains today that divides the Sunnis from the Shiites regarding who was the valid successor to Muhammad. Extensive tribal infighting in various guises continued after the initial unity provided by Muhammad, who died in 632 CE. Regardless of the dominant religion, tribalism continued under Islam, as it had under the Romans, either as pagans or Christians. The Arabs thus lost control of their empire to other tribes who had converted to Islamic faith. These warrior tribes struck out of central Asia. The Ottoman Turks, who proved the most successful, remained the main politically unifying force in Islam until they collapsed in the aftermath of their defeat in World War I.

The Arabs nevertheless were far more politically unified than were the Christians of western Europe during the Dark Ages. Arabic became the common language not only of the holy Koran but of the common people. Moreover, Islam, like Judaism, prescribed detailed rules for daily life. The term Islam itself means "submission to the will of God," and submission helped create the stable cultural ecosystem that emerged under Arab rule.

Stable cultural ecosystems, however, do not encourage innovation and especially not if despotic. That general constraint was reinforced by certain passages from the Koran, according to Karen Armstrong, that cast innovation in a bad light. As we explain next in Part III, Islam's cultural constraints on innovation in a few hundred years render all of Islam highly vulnerable to a newly innovative Europe. In fact, western Europe would emerge from its Dark Age status as a cultural backwater to become the first civilization ever to dominate the entire globe.

As Rome's central authority vanished, local centers emerged to become new frontiers that gave birth to a new and more powerful civilization based on a cultural surge of new technology. European craftsmen invented this technology which in turn reinvented Europe and the Europeans. It also paved the way by early in the twenty-first century for America to become the globe's only mega-superpower, a tribute to a de facto policy of continuous improvement to its techno-structure.

PART III

FROM DARK AGES TO ENLIGHTENMENT

12

FEUDALISM OPENS NEW FRONTIERS OF TECHNOLOGY

The collapse of Rome's central authority following the bar barian invasions was a classic example of how uncon strained behavior by immigrants in ecological release caused a cultural ecosystem to crash. The collapse of Rome also bears some resemblance to how an organism with a "weak immune system" gets sick and dies if invaded by a virus. Rome's collapse, however, also destroyed many of the defenses it had erected against innovation. For example, slavery as such could no longer be sustained since the barbarians, once the main source of new slaves when taken prisoner, had taken control. Although their control was localized geographically, it demolished Rome's central political authority in western Europe. No power since has been able to reestablish central control despite a number of attempts.

Europe's new fragmented political arena gave rise to feudalism. First, Rome's technology and its accompanying software began to atrophy, and much of Rome's practical knowledge was lost. Except in the Catholic church, literacy all but vanished. Most of Rome's highly developed infrastructure such as aqueducts and roads, buildings and monuments began to fall into ruins. Trade all but collapsed after 600 CE. By 750 CE, the once most advanced and centralized civilization on earth became a lowly fragmented backwater mainly controlled by local tribes and former invaders. Western Europe hardly merited the term "civilization" by about 700 CE as cities mostly all but vanished. The civilizations of Islam, India, and China were far more advanced. Still this sad state laid the groundwork for a surge of new technology that would later eclipse them all.

The end of Rome's centralized despotism, however, did not entail the rise of a new democracy in the feudal manors where most western Europeans now lived. These manors were merely local despotisms, not centers of innovation. The villages located between the feudal manors were where the seeds of a freedom to innovate new technology began to sprout. Many of these villages did not come under a feudal manor's control nor a central political authority and by default became largely self-governing. Merchants, free craftsmen, and artisans settled along rivers or other types of crossroads, thus expanding some of these villages into towns. The townspeople supplied the manors with goods the manor owners did not or could not make for themselves in exchange for the food these rising towns did not grow themselves. As the division of labor became more pronounced, these towns began trading with each other. A few towns, such as Scarborough, England, became famous for their fairs that attracted people from the manors and other villages and towns. Knights, second only to manor lords in the social order, came to joust with each other while craftsmen, merchants, and farmers exchanged goods.

In these emerging towns, merchants and craftsmen no longer cowered at the lower end of Rome's caste system. Social stratification continued but mainly in the rural manors, with lords and ladies, knights, freedmen, and serfs. Townspeople had servants but no slaves and did much of their own work. Most craftsmen were free men and not beholden to any master, except perhaps while learning the trade as an apprentice. If a craftsman invented something new, he was free to exploit it (provided it did not upset the local ecosystem of his own craft guild).

Merchants, to be sure, remained a morally suspect group as far as the Catholic church was concerned. The lords of the manor regarded merchants as their inferiors. They ignored the knight's warrior ethic, but the lords had little authority over merchants. Meanwhile, merchant-craftsmen were increasingly admired and rose to control the towns. In fact, Venice, the best run and most prosperous city in the Middle Ages, was quasi-democratic, and officiated by its merchants. According to Jacques Barzun, Venice was closest to an ideal city-state since Plato wrote his Republic.[1]

By about 1200 CE, with a network of towns and a few cities across western Europe north of the Alps, innovations began to emerge. These included mechanical clocks, spectacles, the compass, better ways of weaving cloth, paper and ink, and square-rigged sailing vessels built to North Sea conditions and that did not require galley slaves to row as had the Roman ships. The arts were transformed as well. Brunelleschi developed visual perspective in painting in 1415. Along with such other innovations as new oils and canvas brought painting to life. All that made a big impact. According to Jacques Barzun, this new approach to art was to propel artists into a high status group admired even by the landed aristocracy and the emerging royalty.[2] He also points out such painters had been mere "artisans" dirty with paint smudge and dust before they introduced perspective with its greater realism into their art. Yet after perspective, oils, and canvas, they became "artists." They were admired by wealthy patrons who sought after them in a way unknown in the ancient world. But these late medieval artists, mainly in Italy at first, were also crafty businessmen. They owned and personally managed busy workshops. They often had helpers and a concern for the "bottom line." They were much more bourgeois than bohemian. (The artist as bohemian came much later.)

By 1200 CE, northern Europe's towns became prominent and the pace of innovation was picked up smartly. So was trade. Along the North Sea, the Hanseatic League of towns began robust trading activities in textiles and other products. Their ships were designed to handle the stormy conditions of the North Sea rather than the much more tranquil Mediterranean Sea. These ships thus became the model for the later ocean-going ships that would circumnavigate the globe by the mid 1500s.

In Italy, Genoa and Venice became prosperous and wealthy through trade. Their traveling merchants often brought back new technology from Asia and the Middle East. Perhaps the most notable example was Marco Polo of Venice. He traveled to China, stayed for about twenty years, became an advisor to emperor Kublai Khan of China, then returned to Venice. He brought back Chinese noodles that Italians soon expanded into various kinds of pasta. But the most important items he brought from China were block printing and an explosive mixture of saltpeter, charcoal, and sulfur that was to be called gunpowder. In a little more than one hundred years, both of these imports would, in the hands of free craftsmen, result in innovations that transformed Europe.

China had never used the explosive mixture they had invented for more than fireworks displays. But, thanks to Europe's preexisting metallurgy in making brass church bells, some craftsman decided to elongate and reinforce a bell, drill a hole in the breach, stuff explosive mixture in it, ram a suitably sized stone down the barrel, light it, and *bang*. The stone flew out at great speed and landed a few hundred yards away with a mighty impact. Cannons were used in 1324 at the siege of Metz, France. Thus, Europe gave birth to artillery as a practical use for gunpowder in the fourteenth century. Why didn't China? Perhaps because China had no church bells to suggest the idea of a cannon barrel.

China also invented block printing and just as important, paper, about 105 CE. China's paper had migrated to India and then Islam and had found its way to Europe by about 1200 CE. In 1453, German Johannes Gutenberg took the idea of static block printing and converted it to movable metallic type. Together, paper, ink, and movable type transformed the world through the printing press. No previous innovation in history spread so quickly and transformed so much in such a short period of time as did the printing press. Europe had no printed books in 1455. By 1500, according to James Burke, Europe had about twenty million printed books.[3] They were mostly Bibles to start with but that would soon change. A new industry sprang into being, an early seed of the industrial revolution that came soon after. As Barzun put it, "Gutenberg's moveable type ...was the physical instrument that tore [feudal Europe in] the West asunder."[4]

The printing press marked the beginning of the Renaissance and was perhaps the first example of mechanized mass production. Such presses made it possible to store vast amounts of easily accessible information at so low a cost as to be available to nearly anyone. That fact made future mutations for a larger brain to store more memory largely irrelevant. Possibly, our ability to memorize large amounts of information has regressed as a result of printing.

If any one artifact illustrates the divide between the modern age and the age of feudalism, it would be the printing press. It was to play a major role in the creation of the modern nation-state, the Protestant Reformation, and the rise of mass literacy. It made local European languages, such as English, French, German, and Spanish respectable for the first time because they could now appear in print. An enormous outpouring of literature written in those local languages became accessible to the average person. Newspapers and magazines would soon follow. Indeed, printing in the local language also created a sense of unity such that for the first time people began to identify themselves in terms of the nation where their language prevailed rather than a geographic or tribal identity. That new sense of identity led to the beginning of the modern nation-state as we know it. For example, the King James Version of the Holy Bible was the first version written in English. Thereafter, it played a major role in making Britain's United Kingdom a practical possibility.

Moreover, it is unlikely that German theologian Martin Luther could have made much impact had not printing been available for distributing his ninety-five theses (posted on All Saints Church in Wittenberg on October 31, 1517). Printed copies spread all over Europe in only a few weeks. Many priests before Luther had complained about what many people had felt was corruption in the Catholic church. Mostly these early whistle-blowers had been silenced because no way existed to distribute their views widely. Indeed, even Luther had not at first been motivated by a desire to break away from Catholicism. But he gave voice, through printing, to the frustrations of many.

A change in cultural stratification was directly related to printing itself. Before printing, the only access anyone outside the clergy had to the Gospels or Old Testament was in church liturgy. Each mass or service took certain passages from the Bible as a lesson and then built a sermon around them. Over time, in theory, the faithful who regularly attended church would progressively be exposed to the content of the whole Bible. That strategy was dictated by the fact that reproducing Bibles was very expensive, about one man-year of time. With printing, the cost fell by several orders of magnitude so almost every family could afford to own a Bible. People could thus read and make their own interpretation of a biblical passage rather than depending on a priest to do it for them. Many people felt they no longer were compelled to learn from the priest. Additionally, there was corruption in the Catholic church. Protestanism evolved over the next hundred years and spread across much of Europe.

The printing press, in other words, created a new cultural frontier by disrupting the cultural ecosystem of Catholic Europe. It was FROCA in cultural action. A host of opportunities arose on this new frontier of information technology, because it provided ecological release from previous cultural constraints. One was no longer forced to write only in Latin, which had to be learned with laborious study. One could now write and study in one's native tongue. Moreover, one could now write about theology, politics, philosophy, or the growing field of science. One could also give voice to poetry, plays, and dramatic stories that could be read by many people and not only a few elite speakers of Latin. Printing made a profound addition to cultural information technology and its associated software, and since then, such technology has never seen a major crash.

Cuneiform style of writing gave rise to ancient civilization beginning about five thousand years ago. That was followed by a long period of stasis until the alphabet in 700 BCE, which in turn opened the door to the high point of those civilizations in the west as represented by Rome. It was the alphabet that created new frontiers for science, philosophy, literature, and religious scripture that became prominent in Greco–Roman high civilization. Printing, the third great innovation in information technology, came about in 1453 or about 2,100 years after the alphabet. But unlike the collapse of Rome, after which Roman information technology went backward for a while, nothing like that has happened since. Rather, printing was followed by various forms of electronic information technology. Electronic information technology in its turn has created a host of new frontiers such as telegraphy, radio, telephone, followed by television and most recently the electronic computer, and the World Wide Web.

Note this multifaceted surge in western Europe followed the invention of printing by Gutenberg, a comparatively humble silversmith by trade. Similar craftsmen were to create other such innovations not much later. One of them, the telescope, invented by a Dutch lens maker, Hans Lippershey, would in the hands of Galileo Gallilei early in the seventeenth century open up a new frontier to modern physics. Lippershey's new telescopic lens, when inverted, became a microscope and soon opened up new frontiers in medicine and biology as well as chemistry. Once Galileo turned the telescope on Jupiter and its moons, he was able to confirm the Copernican view that the sun, not the earth, was the center of our solar system. The planets revolved around the sun, not earth, as had long been thought. Later, Isaac Newton devised his three laws of motion and inverse square law of gravity together with a new math called calculus he invented to make the necessary computations involving motion. With Newton's work, modern physics (including astronomy and optics) emerged and transformed the western outlook and its culture. But had not a Dutch craftsmen first invented the telescope, neither Galileo nor Newton could have done their work.[5]

We often think of science as giving birth to new technology, but new technology also gives birth to science. The alphabet was not devised to foster philosophy, theology, religious scripture, or the beginnings of modern science. It was a tool to make possible the translation of business deals from one language to another. Gutenberg knew that printing bibles would be his major market, but his motive was to earn a profit from printing them. The inventor of the telescope had no notion of bringing forth modern physics; he simply wanted to create a tool that let us see more clearly at a distance and thus might prove profitable.

Surging innovations soon lifted western Europe out of its backwater status. Its new technology released Europeans from ancient cultural constraints of time, distance, and power. In about one hundred years, the Europeans—mainly Spanish and Portuguese at first, followed by the Dutch, the British, and the French—had thrust themselves onto the world stage. Europeans explored nearly the entire globe by 1600. As soon as they learned its dimensions, they established colonies and simply took over the weaker, tribal areas. The age of European imperial expansion was to last from 1492 until about 1942, the first year that Asians forcibly ejected Europeans from their colonies. This event was clearly signaled by the fall of Singapore in February of that year.

Europe's imperial expansion and the technology that made it possible had enormous consequences that still reverberate. It triggered the beginnings of worldwide trade, globalization, the industrial revolution, and many wars, including two world wars, a cold war, currently the Iraq War.

13

EUROPE REVIVES TO EXPLOIT COLONIAL FRONTIERS

Three primary technologies allowed Europeans to impose their dominance on the rest of the world. They were: the compass and other instruments of navigation, the square-rigged sailing vessels with stern post rudders that required no oarsmen, and artillery and associated small arms propelled by gunpowder. This package enabled European ships to sail long distances across oceans out of sight of land and yet keep some rough track of their positions. When they arrived at their destination, artillery and fire arms gave them the means to shock and awe anyone who might resist their arrival. In the New World, however, the Spanish had two additional weapons unknown to the Amerindians, namely horses and the virus smallpox.

The Portuguese began the explorations that would end in European colonialism under the auspices of Prince Henry the Navigator. Prince Henry sent ships south to explore the coast of Africa, and the first reached India in 1498 under Vasco da Gama. To the west, Christopher Columbus is credited with discovering the New World in 1492 although Norwegian Vikings had explored there centuries before. Rarely has so great an area been subdued so quickly by so few. The two great empires of the New World—the Aztecs in Mexico and the Incas in Peru—were quickly subdued by relatively tiny numbers. Francisco Pizarro, after invading Peru and with less than two hundred men and sixty-two horses, subdued nearly eighty thousand Incas in a single morning at Cajamarca on November 16, 1532.[1] Earlier, Hernando Cortez had taken over the Aztec empire in Mexico with the fall of King Montezuma II in 1519. Both Cortez and Pizarro, however, were helped by the already established enemies the Aztecs and Incas had created by their own policies.

As discussed earlier, smallpox was unwittingly the single most potent "weapon" the Spanish had. The Amerindians had no immunity to smallpox and similar crowd diseases because they had no domesticated animals. All the major candidates for domestication except for dogs had been wiped out twelve thousand years earlier by the Clovis people. Such diseases originate with domesticated animals or, as with the recent outbreak of SARS, with domesticated birds. After a few mutations, the animal virus can migrate to humans. After the weakest humans die off, the survivors develop antibodies that give at least partial immunity. In America, the native population was nearly wiped out by smallpox before the survivors could develop such antibodies.

Spain plundered immense quantities of silver and gold from the Incas and Aztecs. Then they opened rich silver mines using the natives as slaves in the despotic tradition. Most of the early Spanish colonists, meanwhile, did not come to colonize but to make a fortune, and if they did, to return to Spain. In short, they came to exploit. That ethic of exploitation created a "founder effect" that shaped the later history and cultural ecology of Spain's American colonies. That founder effect proved to have strikingly different consequences from that of Great Britain's North American colonies, which continues to resonate today.

One difference concerns technology. In nearly all of Latin America, the Amerindians who survived the disease epidemics were put to work as quasi-slaves or serfs in mines and then in agricultural plantations growing sugar, tobacco, or similar crops. Thus, much of the ethic of ancient despotic software continued to hold sway. Spanish and Portuguese invaders in South and Central America saw little point in inventing labor saving devices for much the same reasons as the Romans. As had the ancients, they too held physical labor in low regard. Latin America in effect largely opted out of the wave of innovations that was rising in Europe and then quickly took hold above the Rio Grande River. Ever since, few significant innovations or major patents have come from any part of Latin America.

By way of contrast, the first British colonists to New England came to start a new life and to build a new nation based on a puritanical branch of Christian Protestantism. Specifically, these colonists followed the Puritan views of John Calvin. Calvinists did not demean the value of work and indeed extolled its virtues beyond even what the Benedictine Order of the Catholic Church had done with its motto, *ora et labora* or "prayer and work." The ideal puritan would find a vocation and work hard in it. That ideal puritan would also live a frugal and pious life, and avoid ostentation and luxury. He and she regarded saving money as a virtue. Nearly any legal calling would serve as a virtuous occupation. Merchants, artisans, and founders of the emerging factories were equally virtuous in the eyes of God along with preaching or teaching.[2] This puritan or Calvinistic view was clearly a made-to-order religion for the townspeople as well as the farmers of an emerging modern Europe. For Calvinists, this work ethic deposed whatever remained of the ancient view that honest physical labor was socially demeaning. The aristocracy might cling to that view, to be sure. But the landed aristocrats were beginning to lose power in the emergent nation-states to a more puritanical and bourgeois middle class. The new class was made up of merchants, manufacturers, professionals, and artisans who would increasingly call themselves "businessmen." Collectively, they came to be called by the French term, *bourgeoisie*. The term generally refers to middle class people connected with business and its associated professions, who in addition to holding a work ethic, also believe in competition.

The elite of the aristocracy and many of the academics who still clung to the values of Scholasticism (the philosophy of the medieval theologians that propounded the divorce of speculation from observation and practice) did not value the bourgeois who were often dismissed as little more than barbarians who lacked any meaningful cultural values such as an appreciation for music and the fine arts. The bourgeois were also stereotyped as stingy and narrow-minded "grad-grinds" who lacked the ability to enjoy life.[3] ("A Puritan is one who lives in fear that someone somewhere may actually be having fun.") Although it was dismissed and stereotyped, the puritan work ethic played a vital role as part of the software that promoted capitalism and its continuing lust for innovation. It became the founding ethic of New England. After the Civil War, the work ethic set the tone for all the United States, although it did lose much of its original Calvinist theology along the way.

Meanwhile, the imperial expansion and the founding of new colonies had several motives and the first of these was trade. Prince Henry's explorations around Africa and Columbus' voyage to the North America aimed to find a trade route to the Far East where most of the world's spices were grown. Spices, in the days before canning or refrigeration, played a major role in preserving food. But trade routes through the Middle East were blocked by Islamic governments. In any event such trade had long been in the hands of Venice and Genoa. After Columbus, Spain, Portugal, Holland, Britain and France accounted for most of the colonization and global imperial expansion. New technology opened for them a variety of frontiers and a major ecological release from existing political constraints on trade. Technological advancement also provided some release from overcrowding as populations began growing. In the case of the English Puritans, there was also ecological release from the Catholic counter-reformation and its strong and often brutal backlash to Calvinism. The counter-reformation succeeded in holding southern Europe to the church. Italy, Spain, Portugal, Bavaria, and most of France would remain in or return to the church. That success partly accounts for the lack of Protestant interest in colonizing Latin America.

Despite the distance, Spain, Portugal, Holland, France, and Britain had a remarkable success in colonizing India and the

Far East. Britain in the end would hold most of India. Holland took the "Spice Islands" or what is now Indonesia. Britain also took over Ceylon, Malaya, Burma, and parts of Borneo, along with the continent of Australia and the large islands of New Zealand. France dominated Indo China and a few other islands. Both Britain and Portugal established coastal enclaves in China at Hong Kong and Macao. Spain took over the Philippines, and Portugal ruled Timor. None of these areas had strong central governments. They were tribal for the most part or otherwise fragmented. Australia was in the hunter–gatherer stage of human development. All would soon become colonies in the North American pattern, destined to become new nations settled at first almost entirely by Protestant Europeans who came to build new nations.

The internal tensions triggered by Britain's "Enclosure Movement" (1450-1815) also experienced some release from the new colonial frontiers. This movement did much to improve productivity of farmland and at the same time took large amounts of land out of crops–displacing tenant farmers–and enclosed it to form pastures to raise sheep, which increased the supply of wool for the emerging textile mills. The shift from growing crops to herding sheep was motivated by Europe's Little Ice Age. The colder temperatures increased the demand for warm woolen clothing. The higher demand drove up the price of wool. Rising wool prices thus encouraged the large landowners to enclose their farmlands and raise sheep instead of crops. Landowners with good pasture land found that herding sheep was more profitable not only because of high wool prices but because it required little human labor compared to growing crops.

Former tenant farmers had nowhere to go and no prospect of work elsewhere but nevertheless flocked to the cities and large towns. Lacking work, many fell into a life of crime and drunkenness. Families broke up, and their babies–often malnourished and suffering from fetal alcohol syndrome–were abandoned to local parish orphanages. The boys who were raised in such understaffed and underfunded orphanages were released to the streets at age fifteen, often rebellious and full of resentment. They too turned often to crime, and before long, British jails were overstuffed with prisoners.

In desperation, the authorities began negotiating with plantation owners in the southern colonies of North America who grew tobacco. Because tobacco was a labor-intensive crop, plantations needed large amounts of cheap labor and British prison authorities began to sell off prisoners as indentured servants to the plantation owners. In theory the indentured servants could earn their freedom after seven years of service and good behavior. Bad behavior, however, resulted in extended terms of servitude. Later slaves from Africa provided even cheaper labor and had no civil rights attached to lifelong servitude. After black slaves took over the American plantations, the British "transported" their prisoners to Australia where they became the nucleus of Australia's first settlers.

I have written elsewhere about the broader economic consequences of the Enclosure Movement.[4] Here I will limit my discussion to its effect on colonization. In the United States such servitude effectively created a new and second founder's effect in the United States.[5] The South soon developed an ethic (and software) not unlike that of Latin America, and in both regions plantation agriculture played a major role. The prevailing ethic of Europe's landed aristocracy took hold. While American Southerners were mostly Protestant, the leaders were also mostly Anglican, the branch of Protestantism least affected by Calvin's Puritan work ethic. Anglicans in general did not stress the dignity of all work. Indeed, the Enclosure Movement had harnessed the working class in Britain (largely made up of rural refugees who were the world's first proletariat) with almost no social standing. They were generally held by the British middle classes to be untrustworthy, lazy, and prone to theft, stereotypes with some justification given the social demoralization that Enclosure induced.

Thus, the plantation owners in the South, growing tobacco at first, and then later cotton, retained the view that "decent people" did not work with their hands. Instead of an ethic of hard work, frugality, and competitive success that drove the North, Southerners retained the more traditional and aristocratic values of honor, duty, and obligation. These values also entailed a perceived natural right for the elite to enjoy the good life made possible by slaves who did most of the physical work. This sharp division between North and South ultimately triggered tensions about basic values so severe that they led to civil war.

The North's work ethic thus placed the impetus toward innovation mainly in the North[6] and it remained in the punctuation stage of evolution with an ongoing series of new frontiers and opportunities based on self-accelerating technology innovation. The South for its part resisted change, and it struggled desperately to remain in equilibrium.

Meanwhile, Europe's seagoing innovations opened up a new frontier of piracy. As Spanish galleons brought back gold and silver from the Incas and Aztecs, English and other ship owners saw a predatory opportunity to seize these galleons at sea. Sir Francis Drake, who helped found the Royal Navy, began as an English pirate. According to Debra L. Spar, the distinction between naval vessels, merchant ships, and pirate ships was not always clear.[7] For about two hundred years, piracy was well established in international waters and on the high seas, and pirates raided ports on occasion as well, especially those in the Spanish-held Caribbean. But as the colonies grew bigger and the volume of honest trade increased, the demand to stop the pirates became more insistent. The Royal Navy, practically born in piracy under Queen Elizabeth, was by then the most powerful fleet afloat with many interests to protect. The Royal Navy had put a stop to piracy in the open ocean, much as Rome's navy stopped it on the Mediterranean Sea. As such, piracy became a complete FROCA process in itself. New ocean-going technology created a frontier on the high seas without any effective authority. That absence provided pirates ecological release from police authority thus to create a host of opportunities for sea-going predation. Once again, a vacuum of authority led to anarchy and violence. Then too, as so often happens, the pirates overexploited their opportunities and the pirates' predatory frontier closed. Many a pirate ship had been sunk under the guns of the Royal Navy by 1750. Some of the pirates who survived switched occupations and became part of the merchant marine or town merchants. Some became slave traders, but that frontier was later shut down by the same Royal Navy in the following century. After piracy crashed, equilibrium emerged in the new higher-tech sea-going ecosystem based on international trade and commerce.

In the two hundred years from 1500 to 1700 Europe and its new techno-structure had come to dominate the entire globe, an achievement no previous civilization could have done. By early in the nineteenth century, Britain had established the first true worldwide empire with a major presence on all continents and its Royal Navy ruling the seas. International waters had become safe for a burgeoning world trade. That trade, again for the first time, was beginning to integrate the thousands of separate economic ecosystems into a single interdependent entity, and for the first time, into a global economy.

What is equally important, the same technology that empowered Europe's new global reach also created software that launched a new political philosophy. For the first time since the dawn of civilization, this new software and its philosophy challenged and then overcame civilization's traditional despotism based on some variation of the "divine right of kings." In more and more nations, people ceased to live as subjects of the sovereign. Instead they became citizens who collectively were autonomous.

14

DEMOCRACY CHALLENGES DESPOTISM

An activist group from Eugene, Oregon, in television interviews during Seattle's World Trade Organization riots in November 1999, promoted a romantic view of anarchy based on farming. They portrayed subsistence farm families as living side by side in peaceful anarchy. In their own minds the activists abolished the eruption of family feuds, territorial disputes, competition for scarce resources, and other such nastiness. They concluded humans had no need of government.

Alas this fantasy has never existed since hunter–gatherers became farmers. True, humanity survived in small groups as nomadic hunter–gatherers without government. In such circumstances government had no purpose and humans could not support it in any event. Based on Stone Age technology, however, egalitarian lifestyle could support about five or six million people. This is about one tenth of one percent of the world's present population. When over-hunting forced humans to turn to agriculture, they found they could accumulate surplus food as storable wealth. That surplus made it possible to practice the division of labor, and the rapid rise of population thereafter doomed any form of romantic and peaceful anarchy. Humans faced a choice of liberty with murderous anarchy or despotism with law and order. In effect, our forebears gradually handed power over to despots to bring about and maintain law and order, surrendering individual freedom to an absolute sovereign.

But while despotism brought law and order of sorts, it also brought tyranny and kleptocracy in its wake. Despots soon discovered the personal ecological release that comes with absolute arbitrary power. They found they could do nearly anything they wanted. Despots had been charged with maintaining public welfare and a reasonable distribution of wealth although the distribution needed not in any way be related to the contributions made by the people who created the wealth. Indeed, the despots discovered slaves could be highly productive and yet receive only enough in return to survive at a subsistence level. Under the rule of despots, slaves often comprised as much as half of the population and lived mostly in abject poverty whereas the despots lived in opulent luxury.

The Iraq War that began in 2003 between Iraq and the United States and its allies revealed Saddam Hussein as a classic kleptocrat, hoarding wealth for himself and his family. While the Iraqi people often went hungry, he was building luxuriously appointed palaces and stashing away hundreds of millions of U.S. dollars in cash. While Saddam was sleeping in new quarters every night, many of his people feared and hated the despotism and grew tired of his tyranny. But for good reason they also fear anarchy. Iraq again serves as a recent example of what happens when law and order suddenly vanish. Anarchy, looting, and chaos promptly broke out when Saddam's regime vanished. That has happened repeatedly throughout several millennia of cultural evolution. A question arises: is there no middle ground?

In the seventeenth century, and for the first time since the fall of Rome, a serious debate about a middle ground called democracy began. That discussion might not have occurred without the earlier invention of the printing press; major debates about social change usually reflect some new technology's need for new software to go with it.

Ancient discussions such as Plato's in *The Republic* had dismissed democracy as a tyranny by the unjust mob; let's call it "mobocracy." Plato believed Athenian democracy (of sorts–it never applied to slaves) undercut Athens's power by allowing a despotic but well-disciplined Sparta to inflict a defeat on Athens from which the city never recovered. Plato never recovered from the shame of that defeat and in his writing, *The Republic,* he proposed a government by just and reasonable "philosopher kings." Such kings would be trained to rule in the best interests of the whole state and they would have absolute authority. Plato's most famous student, Aristotle, in effect dismissed Plato's scheme as replacing mobocracy with "snobocracy." Aristotle wanted political authority to rest in the middle class that had a personal stake in maintaining law and order and avoiding all extremes. (Neither philosopher proposed getting rid of slavery; both agreed manual work was demeaning and not a fit activity for free citizens.)

In eighteenth-century Britain, the rising bourgeois class of businessmen felt some attraction to Aristotle's notion of authority resting in the socioeconomic center. But they also feared the tyranny of the mobs as another form of anarchy. They had good reason. For example, when King Louis XVI was deposed in the French Revolution in 1793, mobocracy soon broke out in Paris and a reign of anarchy and terror began.

Philosophers Thomas Hobbes, Edmund Burke, John Locke, David Hume, Montesquieu, Rousseau and many others entered into this debate. Hobbes apart, they all wanted to eliminate despotism. They all rejected the divine right of kings. They wanted to create some form of democracy that could ensure law and order and be responsive to the welfare of the people and avoid the tyranny of the mob.

The coming Modern Age, however, gave their arguments a new force. A market economy created by middle-class businessmen was gradually emerging and showed clear signs of self-regulation as political economists Adam Smith, J.B. Say, and others pointed out. Trade avoided zero-sum assumptions where the gains of one participant necessarily come at another's expense. Because of limited resources, zero-sum was more or less true in ancient times thus giving private wealth a bad name in moral terms. But the new technology coming out of the private sector in the late feudal age changed that equation. Craftsmen's inventions such as the printing press, clocks, telescopes, and guns clearly improved nearly everyone's lot. These inventions also opened up new opportunities abroad as Europeans immigrated to America and started new lives in self-governed communities, free of Europe's old constraints of social ranks and obligations.

Britain gradually worked its way into a democracy that used constitutional safeguards to avoid the risk of mobocracy. France had tried but failed, then reverted to a secular despotism under Napoleon. No other major European nation attempted democracy except perhaps the Swiss. They crafted a new federal arrangement that allowed German, Italian, and French areas to merge into a single nation where each area preserved its own language and culture.

Still, it was in Britain's American colonies that a truly new democratic republic emerged in 1776 after breaking away from British rule. Once the colonies obtained their independence, the new nation created a constitution; its philosophy came directly from those earlier debates in Europe. The United States of America was secular from the outset and prohibited from creating any state-sponsored religion. The new government also allowed people to practice a religion of their own choice and believe what they chose. The Constitution also created three separate and independent branches of government; one each for the executive, legislative, and the judicial functions. Moreover, the war of independence convinced most Americans that a confederation did not work because it was too difficult to make decisions in the best interests of the whole. The new central government had strong but limited powers, mainly for foreign affairs and national defense plus the post office. The individual states that comprised the United States had nearly complete local authority. These states, moreover, were divided into counties and cities in which each entity was independent, with its own elected officials, and in which authority was actually exerted over people who in turn had the power to vote the officials in or out of office. Police power also was located at this local level, with each county and city having its own police force.

America's new political package was built on much that was already in place among the settlers who had practiced local self-rule for about one-hundred fifty years. In the North, the settlers lived without a stratified social order. They were immigrants or their descendents who had self-separated from their tribal or other ethnic communities in Europe and thus had experienced ecological release from the social commitments and constraints of the communities they left. This separation led colonists in America to become individualists and focus on self-reliance. They took the idea of personal rights and personal freedom much further than was possible in the Europe they left behind. There, the constraints of communal obligation, while weaker than before, still remained a potent constraint. Moreover, those who remained in Europe were more likely to dismiss an ethic of individualism as little more than a rationale for selfishness and a vice not a virtue. A philosophical difference began to arise between Europeans in Europe and Europeans who choose to come to America as a land of opportunity.

Ironically, these democratic ideas first found favor in England (as contrasted to Ireland, Scotland, or Wales), the least tribal nation in Europe. Invaders had overrun England time and again throughout the ages. Earliest archaeology confirms that the Celts invaded around 75 BCE and displaced the original Britons throughout the British Isles. The Romans followed n 43 BCE and took over only England and Wales by 60 CE, staying out of Scotland and Ireland. After Rome fell, the Angles, Jutes, Saxons, Danes, and finally, in 1066, the Normans all invaded England and then settled in. Not until about 1300 did a language emerge that could, without prior knowledge or special expertise, be recognized as a prototype of modern English. Early English was a mishmash of influences from all these invaders. The other European languages were also thus influenced but on a much smaller scale. Because of constant immigration by outsiders, tribalism never took hold in modern England, unlike Scotland where the old clans remain alive and influential.

Tribes typically resist central authority imposed by other local tribes or by outside governments, as did the Scots and the

Irish, as well as the Arabians up to the time of Muhammad. But where people identify themselves primarily with their tribe, they usually resist the notion that the individual's personal rights trump their tribal obligations; such individualism is seen as selfish and possibly irresponsible. However, because tribal identity among English settlers did not exist, they were able to see the virtue in an ethic of individualism, and it became part of the America's software.

What set American democracy apart was its Bill of Rights. These rights assured people the specific freedom to innovate in the face of a strong central government dedicated to the preservation of law, order, and stability. The Bill of Rights also prevented America's democracy from degenerating into mobocracy. Meanwhile, from the outset America was seen to be a geographic frontier that expanded over time beyond the Appalachians once the Erie Canal opened the Midwest to settlers about 1820. The canal was soon followed by the railroad that opened up the nation to Europeans from coast to coast by 1870. As Frederick Jackson Turner pointed out, however, America's geographic frontier had reached its limits by about 1890.[1] Still, the closing of the western frontier did not close the frontier ethic in America because of a continuing wave of new technology, such as the railroads and telegraph. Without these, the United States might well have remained a midsized nation tied to the Atlantic Seaboard. Once the railroads had built their networks from coast to coast the geographic frontier did close. Still, more new technology such as the automobile, telephone, radio, electric power, and much else created new frontiers of their own.[2] The U.S. Bill of Rights was the main impetus; it set up many powerful institutional protections both for the innovators who created the technology as well as for the entrepreneurs who brought those innovations to market. But here let me state that the protection and nurture of innovators and entrepreneurs, as such, was not the aim of the Bill of Rights. Still, that protection was perhaps its most important consequence, however unintended.

15

NEW TECHNOLOGIES OPEN INDUSTRIAL FRONTIERS

The Industrial Revolution began about 1750, but the technology that launched it had been evolving since about 1400. The foundation had to come first and included replacement of wood by coal as the primary fuel for smelting the ore to make a better iron used with growing sophistication in metal working. Someone also had to invent the early machines used to spin raw wool into yarn and weave that yarn in woolen textiles. It was these machines that gave rise to the first real factories. All these earlier technologies were among the preconditions that made the industrial revolution a practical prospect.

The Enclosure Movement had also emerged and played a major role in bringing about the industrial revolution specifically in Britain. That movement helped rationalize and improve the productivity of farming even as it also increased the supply of wool. It was indeed the textile industry that created the first factories that used machine technology to displace the cottage industry in which families used spinning wheels and hand looms to convert raw wool first into yarn and then into cloth. The new factories displaced that kind of rural labor. But even before the cottage industries went under, many tenant farmers had already been displaced.

Their plight became the main focus of a new movement called Marxism, but we will discuss that later. The industrial proletariat was thus formed by a mass of rural refugees forced off the land by Enclosure when the acres they had worked as tenant farmers were enclosed to raise sheep and abolish common lands. The Enclosure Acts passed by Parliament from 1750 to 1850 continued to transform the rural landscape of Britain, increasing the growing numbers of the homeless and often desperate. But this transformation had been going on for some time. For example, Sir Thomas Moore wrote in the late sixteenth century that, "Your sheep that were wont to be so meek and tame, and so small eaters, now as I hear say, become so great devourers and so wild, that they eat up and swallow down the very men themselves. They consume, destroy, and devour whole fields, houses, and cities."

Meanwhile, technology continued to expand in both agriculture and industry. In 1701, Jethro Tull designed a seed drill. A winnowing machine was invented in 1720 and a water-driven thrashing machine in 1732. A man named "Turnip" Townshend developed four-course crop rotation involving root crops, barley, and seed crops such as rye grass, followed by wheat. A building boom in canals began as well and helped reduce the cost of transport from the country to the cities. All of this added to rural exodus to the cities and caused considerable social turmoil.

By the seventeenth century, British prisons had become overstuffed by these rural refugees and their descendents. Many such prisoners were then transported as indentured servants to plantations in the American South to serve in effect, if not in name, as the first plantation slaves. Later, after the black slaves were brought in from Africa, the plantations no longer took indentured servants. But those who became free formed a new kind of social class that had no equivalent in the northern states. Their understandable suspicion of all authority, forged in the furnace of Enclosure, formed the founder effect for a new and lower social class. The middle class, the plantation owners, and even many of their slaves for years referred to this class as "white trash." After World War II, more often they were referred to as "rednecks." More recently, yet another and harsher term has labeled them "trailer trash."[1] Such terms, often wrongly to be sure, imply a class of crude and uncultured people, or more bluntly, untrustworthy slobs. Thus, the legacy of the Enclosure Movement still resonates today, four hundred years later and more than three thousand miles away.

Enclosure helped to create a social psychology that allowed early economists to dehumanize workers in a commodity theory of labor.[2] Economists applied this theory to a labor force that by 1750 had come to be regarded as practically sub-human. Such people were thought to respond only to draconian and despotic styles of management. Such thinking produced its own implicit Theory X style to go with the commodity theory of labor.[3] Theory X assumes that typical workers dislike their jobs, attempt to avoid work, have no ambition, shun responsibility, are not leaders, are self-centered, resist change, and are unintelligent. But in all fairness, by the behavioral standards of the emerging middle class of Calvinistic bourgeois factory owners and managers, the denigrading behavior of the proletarians often provoked the "riff raff" view held by the middle class. One might intellectually understand why such bad behavior could come about. Such understanding would still not make it easier for even compassionate people who lobbied for better conditions actually to live with that behavior. The solution evolved as a sharp class division that segregated workers as much as possible to limit social contact between lower-class workers and the middle classes.

Early capitalism in the eighteenth century created the factories that required much heavy lifting in some jobs, but dumbed down other jobs into monotonous and dull routines that made room for almost no personal creativity except in a few skilled trades. Worse was to come. Early capitalist factories created very unhealthy working conditions. The jobs paid such low wages that people could barely survive and then only if they worked seventy hours a week. This often alienated the workers in the process and the system made them compete with each other. As economic historian Karl Polanyi has pointed out, competition for place was something new in human society and could cause much anguish in smaller communities. Polanyi argued in his 1944 book, *The Great Transformation,* that prior to capitalism, economic relations were embedded in social relations. After capitalism emerged, however, he argued that social relations were increasingly embedded in economic and market relationships, and that was the essence of "the great transformation."[4] Indeed, a life in a closely knit hunter–gatherer band would have been far healthier and both psychologically and socially far more rewarding. Yet, the physical standard of living in their nomadic lives would have remained at a subsistence level.

Socialist proponent Karl Marx made the plight of this rootless proletariat the focal point of his concern. Marx did not blame Enclosure so much as market competition that forced workers who lacked their own means of production to compete with each other for jobs, selling their labor in order to live. He thus advocated the overthrow of capitalism and its replacement with communism to promote class equality. Marx's mission in turn would create deep and dangerous divisions in Western society. Each system, capitalism and communism, became a universalistic ideology of its own—each with its passionate advocates. The contest between the two was the focus of the Cold War that ended only when the communist experiment collapsed in 1991, seventy-four years after it started in 1917. Within a decade of that collapse, however, the critique of capitalism resumed with a new focus and a new force. It shifted from a concern for labor to resisting economic globalization and protecting the environment.

Karl Marx acknowledged that capitalism had created the industrial revolution and he did not quarrel with that creation as such. But now that "our industrial plant is built," he more or less implied, Marx wanted to communize it to bring order and harmony, peace, and justice to mankind. Marx gave credit to new technology in bringing about the industrial revolution, but surprisingly he all but ignored it thereafter. Instead, he implied technology, having created capitalism, would not change. But as the great economist Joseph Schumpeter was to point out not long after Marx died, capitalism evolved and grew through a process of creative destruction, which caused much of the turmoil that tarnished capitalism in its early days. But the same process continued at full tilt to push capitalism to ever higher levels of technology. During the later days of the Cold War it became crystal clear that the communist system based on top-down command and control could not match the innovation in new technology that the free enterprise system brings forth. The communists could copy it and even improve upon it, but the communist system almost precluded innovation. Soviet president Mikhail Gorbachev tried to reform the system but his effort proved futile. The Soviet "Republics" after 1991 began to break away one by one. By 1992, the Soviet Union had collapsed; the Cold War was over.

Meanwhile, what happened to Marx's proletariat? As capitalism's evolution of more sophisticated technology continued in the more advanced nations, the plight of the proletariat was liquidated along with the proletariat itself. (Technology createth and technology taketh away). Machines had taken over nearly all the heavy lifting as well as most of the routine tasks, both in factory and office. Yet instead of creating masses of unemployed, as Marx predicted more than one hundred years ago, new technology shifted most jobs away from processing raw material to the digital processing of information. Moreover, Marx and other critics could not have foreseen how technology also created new consumer goods and services from automobiles and airlines to video games for computers and wireless computer networking. These new industries created many more new jobs than the new machines in old industries displaced. Where thousands of workers had jobs in America and Europe's early industrial years, scores of millions (when including service components such as the airlines and telecommunications) now have much higher paying jobs that come with far shorter hours and vastly better working conditions.

By the second half of the twentieth century, the textile industry gave birth to the Industrial Revolution in its Capitalist "Release 1.0." Its second Release, 2.0, primarily featured the steam engine, which was first used to pump water out of deep coal mines. Pumping increased the available supply of coal and reduced its cost—the same coal that fueled the steam engines. Coal was also coming into use as a household fuel to replace wood from Britain's nearly exhausted forests. Next, the steam engine replaced the water wheel as a source of power in the textile mills. Most economists argue that the railroad made Release 2.0 so evident that many people would come to see it as the "real industrial revolution." Moreover, faster rail transportation made the iron and steel industry a big business. Between 1830 and 1850, railroad networks were built first in Britain and then in Europe and the eastern half of North America. Closely associated with the railroads was the telegraph. This innovation for the first time made possible nearly instant communication over long distances, which in turn made the scheduling and management of railroads far more practical. All such innovations opened up new frontiers both of geography and technology that provided ecological release from prior constraints. Following these releases, frantic exploitation of the opportunities in the fields of transportation and communications began.[5]

By 1850, steam engines also powered growing numbers of merchant ships and naval vessels. Telegraph lines would soon span continents and shortly thereafter begin to cross oceans. An exponential jump in globalizing human cultures, including their technologies, their economies, and their political systems had taken place. By 1880, in principal, although it was not yet practical, people around the world could communicate with each other almost instantly. But it would take another one hundred and twenty years of continuing improvements and even newer technologies to make that possibility cheap enough and sufficiently widespread to put it into daily practice. Still, the first real steps toward the emerging "global village" had been taken without anyone being quite aware of it. With such steps, capitalism's Frontier Release 3.0, the age of advanced steel and steam, began to unfold. A revolution in transportation, communication, and metallurgy had picked up so much speed between 1850 and 1880 that it became evident the most dramatic technological transformation of human life since the dawn of civilization was in the process of sweeping the world and shifting the industrial evolution into high gear. Thanks to the open hearth method of making steel, the steel industry would soon become the hallmark of industrial might. The Soviet despot, Joseph Stalin, took the name "Steel" for himself as symbolic of his own power.

Capitalisim's Release 4.0 included the telephone, wireless radio transmission, autos, and aircraft. It thrust America into the Age of the Consumer with rapidly rising incomes and living standards. By 1900, hardly more than one hundred years after its independence, the United States became the largest industrial power on earth.

Release 5.0 came after World War II as innovators developed television, mainframe computers, commercial jets, petrochemicals, and much else that together dethroned steel as the emblem of industrial power and might. Today, we are in Release 6.0, the Information Age, which many argue has actually created a post-industrial world based on microchips, PCs, scores of electronic gadgets, cell phones, the Internet, the World Wide Web, and the beginnings of an industry called biotech.

Capitalism's last three releases were launched primarily in America, and its process of creative destruction has been the vehicle of America's advance. Each release represents a frontier, with ecological release from old constraints, and a wave of new opportunities for the pioneers on that new frontier. This is the process embedded in "the American Dream" where one can throw off the old constraints that held one down, break loose into a new and better life, and perhaps rise to the top. Many have done so and risen to the top in the manner of Andrew Carnegie in steel, Alexander Graham Bell, the inventor of the telephone, Robert Sarnov of RCA, Igor Sikorsky of Sikorsky Helicopters, Andy Grove of Intel, and more recently scores of East Indians and Chinese in Silicon Valley. America has long been a magnet for "pioneers" who seek either release from old constraints or new opportunities or both. That magnet continues to draw in immigrants from all over the world first attracted the Puritan immigrants to Massachusetts and Plymouth Rock in 1620.

While each of capitalism's six new releases brought forth new creations, they also brought forth destruction of the old. For the new to thrive, the old must die. That is a law of life and of evolution for culture as well as biology. That law, however, does little to prevent anguish felt by workers that capitalism's creations thrusts aside. Thus, capitalism for all its creation and massive improvement in physical standards of living remains highly controversial. Thomas Friedman in his book, *The Lexus and the Olive Tree,* caught much of the flavor of this tension in the context of globalization in the post Cold War period.[6] The Lexus is a metaphor for the pull of new technology, whereas the olive tree stands for the pull of tradition, a now contentious contest between punctuation and equilibrium.

Evolution, after all, is not all "punctuation" and change along new frontiers. That is actually its smaller part. Evolution normally entails longer periods of equilibrium, and stasis within well-balanced ecosystems. The tension between these dual aspects of cultural evolution generates psycho-circuit overload and ongoing political tension. In some ways, the more successful capitalism becomes at creation the more dissention it generates from those who thirst for stability. We will develop this point in later chapters.

In Part IV we continue to emphasize the creative side of capitalism's new frontiers of technology and focus on America's role in this process. In Part V, however, we will shift our focus more to the destructive side and look at the threat that autocatalytic and self-accelerating technology is widely thought to pose to the earth's environment and its global ecosystem.

PART IV

OPENING THE AMERICAN FRONTIER

16

AMERICA'S NEW FRONTIER

Evolutionary biologists have long noted that new immigrants are a major cause of an ecosystem's disruption and thus become the agents that cause their adopted environment to readjust its ecosystem. It does so through what today's ecologists call a founder's effect, which continues to help shape the behavior of later generations, including the behavior of later immigrants from cultures different from those of the founders who are long dead.[1]

The first European immigrants to Massachusetts, the Puritan pilgrims, left Europe in 1620 to escape religious persecution and practice their faith without fear. Although they were neither the first nor last immigrants to arrive in the New World, they established by far the most influential founder effect that survives to this day as a strong, but now largely secular, work ethic.[2]

The Puritan pilgrims, however, might never have survived without the prior arrival of smallpox. Some French or Portuguese fishermen had gone ashore to trade a couple of years earlier and carried the smallpox virus. That disease quickly wiped out nearly all the Amerindians near Plymouth Rock, who had no immunity to it, as it had in Central and South America.[3] The disease killed so fast, it left their villages with ample stocks of seed corn. The pilgrims discovered these supplies, which allowed them to survive their first winter. They had arrived in the New World poorly equipped for the rigors of their new frontier.

After a few years the Puritan immigrants learned to survive in their new setting on their own. The colony attracted more new immigrants, again mostly other Puritans at first. Settlements founded in Rhode Island and elsewhere were largely self-governing under British rule, although that government began to acquire theocratic overtones based on strict Calvinist principles. Still, settlers also learned about self-rule in "town hall democracy."

Meanwhile, Dutch Protestants settled in Manhattan Island, and English Quakers arrived in Pennsylvania. French Huguenots came to Maryland. New England, however, offered no riches and none of the settlers came to "get rich quick." They did come to start a new life in land where religious theology and practical necessity were free to celebrate hard work as virtuous and deserving of dignity. Such virtues were not compromised by previous traditions to the contrary that Europe's upper classes still cherished, especially in Britain. In the new colonies, most people farmed, but craftsmen were well regarded socially, as were merchants. The farming took very hard work. To farm, the land first had to be cleared of the forests, the stumps pulled, and the rocks removed from the rocky soil of New England. It helped relieve the drudgery if one devoutly believed such work had a spiritual as well as a practical point. The puritan work ethic thus set down deep roots. Combined with the individualism that Protestantism also promoted, that ethic would within one hundred fifty years set the new country's national tone, which in turn vaulted America upward to become, by 1991, the most powerful nation on earth.

New England and the Northeast, however, were not the only new frontiers in what was to become the United States of America. In addition, there was an earlier frontier settlement in Jamestown, Virginia, founded by the English, but these English were not Puritans. There was also a small Spanish colony in Florida and a French colony in New Orleans.

The settlers in the Northeast, especially New England, had an acute sense of opening social and political frontiers, not simply a territorial one. This sense emerged as a doctrine of "manifest destiny," which focused on territory, holding that the United

States should expand from the Atlantic to the Pacific Ocean. That expansionist philosophy ultimately turned the United States into a world power.

In the minds of many people promoting manifest destiny was a belief that God had created the United States with a mission to bring to the world a new enlightenment, as much in a political as in a religious sense. This was more often an implicit assumption; the explicit aim was more practical. The aim was to become a powerful living example of a new, more egalitarian way of life governed by democracy rather than monarchy. Not until the twenty-first century would most Americans understand how much resistance this democratic model would meet in older, well-established cultures such as in Iraq and Afghanistan, which have resisted efforts of American political control.

Before Manifest Destiny could become much more than a hollow hope, a whole suite of new technology had first to come into existence. The original colonies had been hemmed in between the Appalachian Mountains and the Atlantic Coast. Indeed before 1776, Britain aimed to keep its American colonies hemmed and leave the rest of the continent in the hands of the Amerindian tribes according to the Proclamation of 1763. Such conservatism reflected the fact that the high cost of transportation and communication made it impractical to occupy and then settle that region. The Mississippi–Missouri–Ohio river system was a highway of sorts, but until the perfection of the steamboat, it was not economical to move goods upstream. Only travel in both directions could make manifest destiny practical. The first step was to create the Erie Canal that in 1825 connected the Hudson River with Lake Erie and bypassed the barrier of the Appalachian range. The steamboat had been improved to the point of being practical for carrying goods up the Mississippi. Soon, settlers flooded into the Ohio River valley. They formed the new states of Ohio, Illinois, Indiana, Michigan, and Missouri, along with Kentucky and Tennessee. Minnesota, Wisconsin, Iowa, Kansas, and Nebraska would shortly follow.

It is important to realize that the innovations making westward expansion practical did more than open new territorial frontiers. The Bill of Rights guaranteed freedom of association, freedom of expression, and freedom of religion, which were all important institutional bulwarks of what had become, almost uniquely in the United States, a permanently open frontier of new technology. The reason they were needed was simple enough. Rare is the innovation that does not infringe upon someone else in one way or another. No one likes this; people reflexively defend against it . . . if they can. Indeed, they will attack the threat as soon as they see it coming . . . if they can. The Bill of Rights provided a defense against such pre-emptive attacks by protecting property rights and patents of innovators. While such pre-emptive strikes were not impossible, they were difficult and expensive to mount. Moreover, in a largely agricultural society, as America was in 1776, innovations could upset greater society as a whole.

Karen Armstrong made this point well in a different setting. Regarding the rise of Islamic fundamentalism after the Mongol invasions of Islam had passed, she wrote,

> No society before our own could afford the constant retraining of personnel and replacement of infrastructure that innovation on this [modern] scale demands. Consequently, in all premodern societies, including that of agrarian Europe, education [and she might have added, law] was designed to preserve what had already been achieved and to put a brake on the ingenuity and curiosity of the individual, which could undermine the stability of the community that had no means of integrating or explaining fresh insights.[4]

The collapse of Rome, as pointed out earlier, negated these constraints in the small self-governed towns and villages. But the principle remained valid. As Europe recovered and began to prosper thanks in good part to the post-feudal innovations, the traditional conservatism remained in force. Farmers to this day tend to retain a politically conservative point of view. The Bill of Rights, however, countered this conservatism in America's new but predominantly agrarian society. It would not stop innovation.

Two of the most famous innovations were both the work of Eli Whitney. His first was to standardize the parts of muskets, thus making it possible to mass produce them, and with that advancement he won a contract from the army. His second was the cotton gin that reduced the labor content and improved the quality of raw cotton made ready for shipment to market. Only after the cotton gin could King Cotton rise to take over the South's economy and ironically create an expanded market for slaves. Cotton distribution was helped by the steamboats that came shortly thereafter.

A whole host of other innovations followed. Samuel Morse had invented the telegraph by 1840, and the railroad boom that followed by 1850 well and truly began transforming the country. The railroads, while opening vast acreage for farmland, also fostered rapid industrialization that pointed to the end of agrarian domination as a way of life. Until about 1860, these innovations could be said mainly to have opened geographical frontiers. They continued to do so for a few more years, but the new business frontiers these innovations opened thereafter rose to prominence.

In the history of the world, the primary role of political government was to achieve stability through disciplined law and order. In the evolution of culture, government continued to focus on achieving and preserving stability and equilibrium and thus almost never was a force for change through innovation. The only frontiers it promoted were territorial and aimed at extending the reach of its own authority and acquisition of resources. Despite the Bill of Rights, more often than not, innovations in technology were viewed suspiciously because they had the potential to disrupt the status quo. A government sometimes promoted a specific invention to solve a specific need, as the British had with the chronometer. The clock-like chronometer was a significant technical advance that enabled mariners by 1772 to calculate the exact difference in their time versus Greenwich Mean Time and thus determine their precise longitude. It was invented by a carpenter by the name of John Harrison. The British elite academic establishment was much aggrieved by this, and for ten years the Board of Longitude delayed paying him the promised financial award for his contribution.

Inadvertently, the American government, thanks to the Bill of Rights, was the first government in history to become a powerful promoter of innovation. It did this by protecting innovators and inventors' products and ideas. Disruption and change became a necessary part of personal freedom, whether through hard work, creative thinking, dumb luck, or whatever, and led people to accumulate much greater than average wealth as an acceptable outcome of their effort. It necessarily followed that great inequalities in the distribution of wealth would arise.

The FROCA process that applies to the new frontiers created by innovations also applies here. The innovation occurs. It disrupts the status quo and creates new frontiers. These frontiers provide ecological release for the pioneers along those frontiers and they try to exploit those opportunities as fast as possible, causing the boom period. But these opportunities are almost always overexploited and a crash follows, causing the bust period. Many of the pioneers become extinct through bankruptcy, and the survivors are then forced to adjust to a new set of constraints.

Risk-taking individuals, who are the pioneers of innovation, are also the kind of people who take license with the old rules they feel no longer apply in new situations. Although this view is often true, taking license usually is carried too far, which is obvious after the crash comes but not always before. After a crash, rules become a major force for maintaining some stability in the new industry. Audits for compliance now begin to take precedence over performance evaluation. Corporate management becomes more bureaucratic and much less entrepreneurial. Concern for process now trumps outcome. The new industry matures. Its capacity to accept further innovation declines sharply because innovation threatens stability in the new ecosystem. Corporate management becomes more like traditional government. It aims mainly to preserve stasis, not promote change. Managers accept change when competition or other circumstances force it upon them, not because they innovate. Debora L. Spar recounts the FROCA process using other terms. One of her main points is that in the early heady days of the frontier, the pioneers all proclaim a new age where the old rules do not apply. That spirit can change quickly, as Spar points out. After the pain and chaos of a crash, the survivors are suddenly willing to write new rules to maintain stability.[5]

Were it not for the Bill of Rights, large and mature businesses could constrain further innovation. They have as their mission "the bottom line," a mission that reflects survival first. Greed comes later. Management by law is charged with preserving their investors's equity. Corporate governance, like political or academic governance, can become kleptocratic and use the system to enrich the company elite. When those at the top take too much for themselves another kind of crash follows: rivals may overtake their market advantage. Competition between rivals can occur even in a market dominated by only a few companies.

What makes America unique is how, after one frontier crashes and closes, new ones continue to open up as the innovations keep coming through the telegraph, telephone, radio, television, and so on, to the internet. In the sixty-year period from 1820 to 1880, the railroads and telegraph let America achieve its "manifest destiny of creating a nation that spanned the continent from coast to coast. Fredrick J. Turner is famous first for citing the closing of the western frontier, but he also pointed out that America's frontier spirit survived that closing."[6] Indeed it did. The American nation's frontiers are now technological, created by innovation. They have been since 1880 sparking still more remarkable transformations. For example, average American real per capita income (net of inflation) has jumped more than six-fold since 1880. That is by far the greatest widely distributed income leap in history.

Before going on with these later transformations, let us first return to the 1860s when the clash between America's two rival cultures came to a head. It ended in a civil war that was, in relation to our population, the worst blood-letting in American history. For our purpose, the importance of the Civil War relates to a clash between rival founder effects on the American territorial frontier. The founder effect in the North favored change, new frontiers, and thus in effect, the punctuation phase of evolution. The founder effect in the South led that region to resist change. The South defended the equilibrium of a way of life led by a "landed aristocracy" that depended fundamentally on slavery.

So to that story we turn next.

17

CLASHING FOUNDER EFFECTS AND THE CIVIL WAR

From its onset, the issue of slavery divided the United States. But until about 1820, when the nation was largely agrarian, slavery did not appear to threaten the union itself. The disjunction between a legally sanctified institution of slavery and a Declaration of Independence that proclaimed God-given rights for all was clear enough. This disjunction was confined to the South because slavery was illegal elsewhere. Moreover, the balance of power in Congress was more or less stable. The North did not have enough votes in the Senate or the House to outlaw slavery and was not prepared to end it by force.

After 1820, however, the issue of slavery became increasingly acrimonious. Several innovations account for this fact. First, the Erie Canal, completed in 1825, opened up the Midwest by linking New York State's Hudson River to Lake Erie. New settlers from New York and New England were against slavery. Second, the railroad and telegraph tended to amplify a trend for creating more new states against slavery than new states for slavery. Perhaps even more important, however, was the fact that these two innovations began to amplify the natural bourgeois tendencies of the North. The industrial revolution was beginning to get underway in the North but not in the South. That revolution would amplify, and be amplified by, the North's more Calvinist tendencies. The railroads, moreover, were being financed out of New York, and not out of Richmond or Atlanta. [1]

Between 1820 and 1860, therefore, the divide between the North and South grew wider on all fronts, practical as well as ideological. Politicians in both regions repeatedly attempted to reach some compromise or accommodation on the issue of slavery. Many agreements were struck only to fall apart later on. For one thing, Eli Whitney's cotton gin combined with the industrial revolution in England had promoted cotton as an ever more important crop in the South, displacing tobacco as the mainstay of the South's plantation economy. It was widely believed in the South that cotton fundamentally depended on slavery. Somewhat ironically, therefore, the industrial revolution that was causing more intolerance of slavery in the North (industrial employees, such as mill workers were horrified at the prospect of their jobs being taken over by slaves), actually made the South more dependent on slavery.

Additionally, new industries arose—iron and steel for example—along with many fabrication plants for rolling stock and steam engines. New England developed its own textile industry based on Samuel Slater's invention of yarn spinning machines. Better guns—the Colt revolver for example—were developed in the North. This industrial boom reinforced the northern bourgeois ethic that was incompatible with the southern tradition-bound ethic reflecting the values of a landed aristocracy, freed of the necessity of manual labor, by a subclass of slaves as in the ancient despotisms. Alex de Tocqueville outlined this difference even before 1840. [2]

Early on, the South hoped to add new slave states as fast as the North added non-slave states. For example, although Maine was admitted to the Union in 1820 as a free state, Missouri was admitted in 1821 as a slave state. Congress abetted this effort by separating North from South with an imaginary geographic boundary. Any further state entering the Union to the south of that line would be a slave state. By circa 1850, it seemed clear that geography favored the North, as most of the Louisiana Purchase from which so many new states later came lay north of the Mason Dixon dividing line. Border states in particular were hotly contested by both sides. Civil war broke out informally in several states years before the main event. Northerners led by John Brown, among others, helped slaves escape from the South and become free citizens in the North, which led to a law Congress passed to stop this practice–a law that aimed to mollify the South. The Compromise of 1850 only inflamed northern anger by allowing capture and return of fugitive slaves and permitting new western states to decide for themselves whether to be free or slave states.

Southerners saw the day coming when slavery would be outvoted in the Senate. They feared that the South would then be forced to abandon not just slavery but a whole way of life and a set of traditions that went with it. Duty and honor, good manners, genteel civility, and leisure characterized by elaborate social functions with grand balls that featured gracious social etiquette all would end if the North's ethic of individual ambition prevailed in Congress and the White House. When Abraham Lincoln won the 1860 election for the new anti-slave Republican Party, the southern states began to secede from the union, starting with South Carolina. Lincoln did not promise to end slavery, but he did not have to do so. Southerners knew that if enough states became non-slave and a Republican was in the White House, slavery would end.

After the South Carolinians fired the first shots at the federal forces at Fort Sumter, in the spring of 1861, the war began. Most of the best officers in the U.S. Army, educated at West Point and trained in the war with Mexico, opted to join the South: Lee, Jackson, Longstreet, Hill, Stuart, Early, and others. They had come from the South, adhered to southern values, and felt more loyalty to their respective states than to the whole nation. They claimed to be fighting, not for slavery, but rather for states rights. Still the "right" they were fighting for was the right of a majority of people in the South to impose slavery on a minority of people denied protection of the Bill of Rights.

The early battles were nearly all resounding southern victories beginning with the Battle of Bull Run in Virginia. If the

South was thin on resources it was long on good generalship, fighting spirit, and high morale. The North had well-trained soldiers, to be sure, but their generals proved to be inept compared to their rivals in the South. They lost battle after battle, twice at Bull Run, several others during the 1862 Peninsula Campaign, at Fredericksburg, the Wilderness, and Chancellorsville, plus many smaller conflicts. Indeed, the only battle the North had clearly won was at Shiloh in Tennessee in April 1862. Another battle in September at Antietam (or Sharpsburg) in Maryland was a tactical draw but proved a strategic northern victory because General Lee's severe losses forced him to cancel his planned incursion into Pennsylvania. In nearly all the federal defeats, meanwhile, the North possessed larger forces and they were better equipped. Still, the North's two successes proved crucial. At Shiloh, General Grant proved his ability (he too was a West Point graduate and veteran of the Mexican War where he had served with Lee). More to the point, Grant used Shiloh as a springboard to conduct a long campaign to cut the South in two. By July 1863, Grant had taken control of the Mississippi River by the capture of Vicksburg, Mississippi.

The North could afford its losses, however, while the South could not. After two major post-Antietam victories, in June 1863, Lee tried again to invade Pennsylvania hoping to capture the federal capital of Washington, D.C., from the rear. Instead, Lee was soundly defeated with terrible losses at Gettysburg in July just as Vicksburg also fell to Grant. After that, the South had almost no chance of winning. The North continued to gain strength, confidence, leadership, and better equipment. All the while, the North showed a continuing capacity to accept huge losses and continue fighting. For example, at the battle of Cold Harbor in Virginia in the summer of 1864, the Army of the Potomac suffered five thousand casualties killed in just twenty minutes of battle. Still, Grant kept hammering at Lee and gradually ground him down.

Meanwhile, General Sherman split the South in two yet again. After he captured Atlanta in late 1864, Sherman launched his "March through Georgia to the Sea." Grant, despite his own heavy losses, finally shattered Lee's army in Virginia in the early spring of 1865. The war ended at Appomattox in April 1865.[3]

Both sides together lost about six hundred thousand out of a combined population of about thirty-two million people. The South fought its cause to the point of utter exhaustion, long after all hope of winning had passed. Yet ordinary southern soldiers, almost none of whom were aristocracy or owned slaves, continued to fight ferociously to the end. By then, many were going barefoot, wore rags, were short of food and medicine, and were without military supplies. Jay Winik argues that even in these dire circumstances, the nation barely escaped a continuing guerrilla war, in part because Grant had offered generous terms to Lee. Besides allowing Lee to keep his sword, his defeated soldiers were permitted to keep their horses and side arms. [4]

In *The Lexus and the Olive Tree,* Thomas L. Friedman used the Lexus as his metaphor for high technology and the olive tree is his metaphor for the pull of tradition. In America's Civil War, the North was fighting for the Lexus, the South for the Olive Tree. The North was pushing for the high-pressure dynamic industrial world characterized by ongoing innovation and change. Still, to gain the maximum benefit from such a society, the North felt it had to preserve the Union.

The South fought for the genteel traditions of a plantation society run by a gracious, cultured, and landed neo-aristocracy. Both sides accepted frightful losses to pursue their vision of their own core self-identity. In terms of evolutionary theory, the North's identity was attached to the Punctuation phase, the change mode, a frontier society. The South's core identity was attached to its established Equilibrium phase. The North wanted to pursue innovation while the South wanted to constrain it. In this case, the North's Lexus of technology had sufficient influence to uproot the South's olive tree of tradition. Both sides were under the influence of autopoiesis, or "self making," a concept explored in more detail in Appendix 2.

The North's victory consolidated the Puritan founder effect for the nation. That effect thus continued and accelerated its dynamic of innovation protected by the Bill of Rights. The South did not participate in America's industrial revolution that followed the war. It had not much participated in industry before the Civil War and would not do so significantly until after World War II, nearly one hundred years later.

By 1869, the Atlantic and Pacific oceans were united by railroad and telegraph, joined together in Utah. The steel, oil, refining chemical, textile, and other industries began dynamic growth that in thirty-five years transformed the United States into the largest industrial power on earth. Still, the real dynamic behind the now purely technological frontier that America represented was only then about to reveal itself. America was poised on the brink of the greatest and most widely distributed leap forward in the average material standard of living ever. By 1900, America had achieved the status of a world power. At that time, it was one nation among many that included Great Britain, France, Germany, Austria-Hungary, and Russia. By 1991, however, none of those nations matched the United States, which stood alone as the world's superpower. That status evolved directly from America's ongoing pursuit of technology, which was protected by the Bill of Rights and sustained by a Puritan founder effect that enshrined such bourgeois values as hard work, competition, and the morality of achieving personal wealth.

18

THE HIGH-TECH FRONTIER TAKES OVER

America may have become the foremost industrial power in the world by 1900, but farmers still made up about a third of the population of seventy-five million people, 60 percent of whom lived in rural areas. Many industrial workers, for example those in sawmills and mines, and railroad and telegraph workers, were spread across the entire nation. Additionally, domestic household help constituted the nation's largest single group of hired employees. No airplane had yet flown. The automobile was still a horseless carriage seen by a comparative handful of people in 1900. The telephone was just beginning to come into homes. Wireless telegraphy (radio) was about to make an impact, but the first message by radio had yet to be sent and received across the Atlantic, and there were no radio stations.

The average standard of living was somewhat higher than it had been but it had not yet escalated. The average farmer in many rural areas could still barely eke out a living; he and his family probably lived little better than had his counterparts in biblical times. Poverty, and often grinding poverty, remained a fact of life for the majority of people and was thought to be more or less inevitable. My own mother at age ten lived with her family for a time in a one-room log cabin with a dirt floor as late as 1914. Society assumed that "the poor will always be with us." Before the rise of labor unions, industrial workers worked long hours in sometimes dirty jobs with miserable conditions for 10 cents per hour. A dollar a day was then considered a standard working man's wage. The average weekly wage, according to Cox and Alm, was $8.88 and, according to the U.S. Bureau of Labor Statistics, was $12.98 for 59 hours per week. Wages did not allow most workers to own their homes.[1]

All this would shortly change. A whole suite of innovations had either come on stream or were about to do so. As a package, they opened, or were about to open a whole series of new technological frontiers over the next twenty years. By 1930, the American nation had been transformed. For the first time an end to poverty as the average condition of life seemed clearly in view. The Great Depression, an early crash of some of this technology, intervened during the thirties, but by 1945 it was up, up, and away again. Yet even during that depressed decade, thanks to continuing improvements in technology, the airlines, radio, and the movie industry were booming. But they were new and hence relatively small.

The auto industry induced the most obvious transformation. Automobiles powered by internal combustion engines fueled by gasoline were invented in Europe where the marriage of this engine to a carriage first took place. But the automobile first became a major and transforming industry in the United States, thanks in good part to Henry Ford's pioneering efforts. His story has been told so many times that we need only hit the high spots here. Ford dropped out of high school, but became a self-educated engineer (of sorts). Early in life Ford became fascinated by the horseless carriage and he finally managed to begin a firm to build them. After several false starts he settled on his Model T. Ford had a vision of building that model in a way that would reduce the cost enough so that nearly any worker could afford to buy one. He conceived the idea of assembly-line manufacturing in 1908. Before, workers moved from car to car as they finished doing their particular job. Ford built a conveyer belt that moved the car along to the workers. The workers stayed in one place doing their particular job, and then the belt moved the car to the next worker, and so on. The labor savings were dramatic. Ford promptly began cutting prices to increase sales.

Suddenly the main constraint was the boredom of the assembly line. At the going rates of pay for factory workers, few stayed if they found a more rewarding job at around $2 per day. Labor turnover rates were high, and the quality of the work suffered as a result. In 1913, Henry Ford introduced a then-radical idea. He more than doubled the rate of pay for the assembly line workers to $5 per day. He reduced the hours a bit as well. In those days, $5 was a decent middle class wage, unheard of for common labor. The results were stunning. While most observers thought Ford had taken leave of his senses, workers nearly rioted in front of Ford's Personnel Office trying to get hired. Suddenly the Ford Motor Company had its choice. The labor turnover rate fell sharply. Quality went up, and Ford sold cars by the thousands month after month, year after year. Ford's own workers were quick to buy what they had produced.

Other manufacturers soon followed Ford's lead. American workers at last began earning a middle-class wage that commanded some buying power. The work itself may have been dull, but at least it paid well. Workers could now enjoy their time off, since they had more of it, and drive their cars into the countryside for picnics with the family. Some relocated from crowded urban slums or ghettos into more open, single family residential neighborhoods on the edge of town. Suburbia was born, as was commuting to work by car. Some claims have been made that attempt to belittle what Ford did. He was not really aiming to help people, the critics allege; he was just greedy for more profits. Besides, he imposed a paternalistic moral code on his workers that aimed to keep them sober, etc. When news of $5 day first got out on a freezing January day, 2,000 workers showed up hoping to get hired at Ford. Of course only a handful actually got hired and they had to endure a six months period of probation to qualify for $5 per day. But probation and paternalism were standard in those days. A $5 day with or without paternalism was a bombshell. Here, however, Ford's motives are irrelevant. The $5-a-day wage had a dramatic effect on the American economy and its culture regardless of Ford's inner motives.

During the same time, the telephone was rapidly being improved and made less expensive and so it too began coming into middle-class homes. Indoor plumbing with hot water, bathtubs, and flush toilets was becoming standard in any new house and most major remodels. So was electrical power, now that it was generated in central power plants. By 1920, all the urban centers and most of the suburbs were electrified. Electrical appliances such as stoves, toasters, smoothing irons, and refrigerators were supplementing simple electric lights in middle-class homes, and before long, in all homes. Electrical street lights replaced gas lights in towns. Motion pictures became popular as mass entertainment when the motion picture industry grew and especially after it settled in Hollywood, California, shortly before 1920. Wages had more than doubled from $700 per year in 1914 to more than $1,500 per year by 1929.

Most of this material progress is well known and well documented. World War I intervened, to be sure, an event we address in the next chapter. In any event, that war did not check the rush of innovations and new technology. By 1930, America had left the horse and buggy era, as it was then called. While the buggy itself was not old, horses or horse-drawn vehicles were the only means of personal transportation for upward of five thousand years. In a mere thirty years, America left that behind.

The automobile, meanwhile, had solved an enormous urban pollution problem. As motor vehicles began to replace animals, the filth and stench of animal manure and urine that polluted most city streets, especially after a rain, began to vanish. Once the transition was complete, for the first time since the dawn of civilization, cities found it possible to keep their streets relatively clean and odor free.

In the first thirty years of the twentieth century, America's culture had been transformed. Many behaviors changed as a result. For example, when the typewriter, the telephone, and the automobile were perfected, they triggered a revolution in manners and morals. The typewriter and telephone opened up new opportunities for the employment of young ladies in business because women had greater finger dexterity than men, typed faster, and did so more accurately. Then too, business people in general preferred to hear a female voice answer the telephone. Without a backward glance young ladies in the thousands began to quit their jobs as domestics to become telephone operators, receptionists, and secretaries, and to do other clerical jobs in business and government. Moreover, while strong cultural pressures from wives inhibited males who employed domestic servants from entering romantic relationships with them, far fewer such inhibitions were found in offices.

In 1900, nearly all secretaries were men. By 1920, they were largely women, thanks to the telephone and typewriter. Bosses soon began to marry their secretaries. Many a well-spoken young lady found that by age twenty-five or so she could leap from lower middle class status into the upper middle class ranks or even higher and do so without guilt or shame. Few young men could do that.

The Victorian moral code began to atrophy. Here the automobile played a major role, abetted by the telephone, and with some help from the typewriter. Given this suite of new technology, a young man could bypass the Victorian parental filtering system that required permission of the parents for him to pay court to their daughter. If such permission was granted, he could then come to call and sit by his girlfriend in the parlor with one or both parents present. But when the telephone became common, he could call directly to make a date. Or if he spotted the object of his affection walking home, he could offer her a ride in his flivver and perhaps take the long way home. "Oh won't you come with me Lucile in my merry Oldsmobile....", as an early tune had it.

In practical terms all this was the beginning of women's liberation. Not only were women suddenly able to date more freely, they were also able to take a respectable job as a secretary or telephone operator. Young ladies began to leave home and perhaps share an apartment with another working girl roommate, discovering at the same time a degree of personal freedom unknown by domestics. Both the pay and the marriage prospects were better. In domestic service, a girl's morals were scrutinized by the matron of the house. Business firms rarely took on that quasi-parental role. For a good account of this process of liberation just after it happened, read Frederick Lewis Allen's 1931 book, *Only Yesterday*. His chapter "Revolution in Manners and Morals" focuses on this liberation. Not only had the horse and buggy era become obsolete with the arrival of new technology, so had much of the Victorian moral code. In the Roaring Twenties, that code had partly been replaced by the age of newly liberated flappers, jazz, bootleg booze, speakeasies, women smoking, and a now more relaxed view of premarital sex. It was seen as an era of "anything goes" as the title of a popular song by Cole Porter put it. Women had left home for the workplace in larger numbers. Once the war ended men were often reluctant to return to the confining mores of American society, and women were less willing to abandon their new freedoms outside their homes. As the then-popular song had it, "How Do You Keep 'Em Down on the Farm After They've Seen Paree?"

This suite of technologies opened new frontiers on many fronts at once. Each frontier offered its own ecological release from the previous cultural constraints. Changing culture created technology, and that technology remade our culture. Massively, people exploited the new opportunities. Many people soon began to overexploit them. Not surprisingly, this boom of the Roaring Twenties ended with a big crash, the Great Depression. I have written extensively about the Great Depression in economic terms in earlier work. Here, I will briefly summarize the main points I raised. My explanation is quite different in some respects from the usual ones on offer.[2]

The widely popular idea that the stock market crash of October 1929 "caused" the Great Depression is overblown. To illustrate, a more dramatic crash on October 19, 1987, did not even cause a recession. Rather, the stock market crash of October 1929 was the bursting of a speculative bubble, caused by the wild overexploitation of a financial innovation that allowed people to buy stock with 10 percent down. If they did so, and if the stock's price rose by 10 percent, the purchaser doubled his money. People who experienced that success once or twice suddenly saw vistas of great wealth open before them. Many people then rushed into the stock market to speculate on margin (the 10 percent down). As they did, the price rose. Within a year, average prices had doubled, encouraging more people to buy. Speculators who saw the price of a stock double, at 10 percent down had reaped a ten-fold gain. For instance, they borrowed $1,000 on personal credit to make the down payment and the stock doubled in price, the speculator made $9,000 after paying back the loan, nearly a ten-fold increase on the investment. He or she would have made that sum without investing any of their own money. Such prospects proved far too tempting to resist.

Thus, the 1929 stock market bust is a good example of a fast-paced FROCA process in action. A financial innovation opened the frontier of leveraged speculation. The pioneers discovered ecological release from the necessity of saving hard-earned money to invest; they could do it on credit using other peoples' money. It was a new opportunity to be sure, but one that was promptly overexploited and was thus followed by significant economic depression. In the adaptation phase that followed, Congress passed laws by which, when combined with the low interest rates imposed by the Federal Reserve Board, credit risk at low rates precluded leveraged speculation until interest rates were raised again about thirty years later. Low interest rates inhibit leveraged speculation by restricting the supply of credit. Without credit, there is no leverage.

One element of the Great Depression itself must be mentioned to illustrate FROCA on the industrial as opposed to the financial side. I use the auto industry to illustrate FROCA because it played a major role in the broader depression. Innovations create new industries if they achieve market acceptance at an early stage. Recall that innovations work in cultures like mutations of DNA work in bodies. The mutation takes hold and propagates to future generations if it provides greater fitness in the ecosystem. Innovations gain acceptance in the market (an ecosystem in itself) if it offers a comparative advantage or solves some problem. If it does, it provides the equivalent of greater fitness in the cultural ecosystem. The automobile did just that.

There are many variations on the theme, but once an innovation takes hold in the marketplace to attract enough investment to call itself a new industry, it usually begins to grow rapidly in percentage terms, although rather slowly in actual numbers of units sold. At high percentage rates of growth, small unit numbers at the beginning mount up. This kind of rapid growth lasts until the level of technology that defines the innovation saturates its market in terms of its fitness. It then matures, growth ceases, and a crash ensues. At maturity, the new industry's output drops from growth plus replacement demand to replacement demand only. The length of time rapid growth continues depends on the improvements made to the original technology. If the improvements come fast, as they did with the automobile, the period of rapid growth can last thirty years or longer. The definition of improvement meanwhile is precise. If it both lowers the cost and improves the technical quality of the new product at the same time, growth is explosive. In the case of the auto industry, the cost of a new car fell by about 80 percent, and its quality improved enormously as enclosed cabs, reliable engines, self starters, headlamps, heaters, windshield wipers, and so on were added one after another. The safe highway speed rose from about 15 mph to about 60 mph on paved roads.[3]

Auto registrations thus rose from about one for every two hundred people in 1909 to about one car for five (or one per family) on the average by 1929. The auto industry became the largest manufacturing employer in the nation. Five million autos were manufactured that year, about half for replacement and the other for growth. In 1930, demand growth ceased for the 1929 level of technology that was temporarily stable. Production at once fell by half to replacement demand. The drop in the demand for motor vehicles meant an equal drop in the demand for and output of steel, glass, rubber, electrical items, paint, and so on. Since labor worked as hired hands paid fixed rates rather than as partners paid rates that depended on the amount of revenue available to pay them, drops of demand resulted in equally sudden layoffs of labor. Unemployment shot up and aggregate income plunged. The demand for autos fell below replacement demand and the nation's economy crashed at the end of thirty years of dynamic growth.[4]

Inability to predict a maturity point means that manufacturers produce as fast as the orders come in. If business is booming, few people focus on serious planning for a sudden collapse in demand. That is equally true today and one need only point to the dot-com crash of 2001–02, for an example.

A similar pattern had taken place earlier in the railroads and telegraph business that matured in the 1890s. All new industries before and since have gone through some variation on this theme of rapid growth (as the new technology is being exploited as fast as possible) to maturity when those opportunities suddenly cease. A crash follows because the level of output implicitly assumed present growth will continue. When it does not, the crash is often as abrupt as it was with the railroads in the 1890s, the auto industry in the 1930s, or the dot.coms in 2001.

Given the large number of such crashes, including those involving leveraged speculation (the fast track both to "get rich quick" or "get poor quick"), one might have thought humans would learn from experience. What has inhibited such learning? In nature, after an organism is born, its growth continues fairly fast to maturity. Then its growth ceases. A similar pattern applies to the growth of new industries. But while we have plenty of data about the growth of life to its maturity point, economists have been tracking new industries' growth for a much shorter time. While we know they will mature, we cannot readily say when. If we misjudge, a crash follows because, however unwittingly, we have overexploited the opportunities the new technology presented when it gave birth to a new industry. The telegraph, railroad, automobile, electrical power, radio, airline, electronics, and many others have gone through this FROCA process and crashed when demand matured and growth slowed sharply.

At this point it might be well to provide an example in more depth concerning the airlines. In fact, it was in this industry that I had my own first tentative inklings of FROCA long before I had any idea of writing this book. My experience began in late 1960 at the Boeing Company. In June of that year, Boeing had hired me as a market research analyst, fresh out of the University of Washington's Graduate School of Business. I had completed all the requirements for my doctorate except for the dissertation although I did have an approved topic: I wanted to discover the cultural values that accounted for Japan's early economic growth and abrupt shift to an industrial economy.

My first major Boeing research assignment was to study the nature of demand for airline traffic that had been growing at a robust 15 percent per year but had suddenly slowed almost to a stop. This sudden slow down was beginning to cause great concern. Boeing had just launched its 727 medium range jet program that had been justified only by a forecast of continued rapid growth. Forecasts throughout the industry, including those of government, all predicted high rates of growth continuing without let-up for at least another ten years. Huge new investments were being made on the strength of these forecasts by airplane manufacturers, the federal government, and others. The federal government was building new airports such as Dulles outside of Washington, DC. The states were upgrading almost all major hubs.

Once growth seemed to stop, however, many analysts decided the "end was near." For example, the chief economist of American Airlines, Dr. George Hitchings, explained the robust postwar air traffic growth as merely an exchange of shares between the railroads and buses on one hand and the airlines on the other. Hitchings pointed out that the total growth of inter-city common carrier traffic had remained flat since war's end. All net travel growth, he maintained, was being carried by the private automobile responding to the new freeway system then under construction.

Boeing's top management was so worried by the logic of Hitchings's argument that the CEO, William Allen, commissioned a study to plan how Boeing might gracefully exit the commercial jet business. The Transport Division was less than enthusiastic about such a plan. The director of sales and marketing asked me to find out if Hitchings's arguments had any weaknesses that we could use to refute it. As the only economist around and freshly out of graduate school, I was given the job. So for the next six months I was sort of sleepless in Seattle, working by day on trying to understand the demand for airline travel and by night trying to understand the role religion played in Japan's abrupt decision in 1868 to become an industrial power. The issue of technology came up in both studies. Japan saw that without Western technology it would become a colonial dependency of the west. It thus launched itself onto a new frontier to catch up in order to preserve Japan's autonomy and at the same time to retain as many of its own traditions as it could.

On the airline side, the puzzling thing was that the sudden halt in traffic came about just as the jets were entering service in large numbers. Everyone thought that people would flock aboard. That they did not startled nearly everyone and came as a shock of sorts. I was a total novice here. I knew little about the demand for air travel beyond that of a layman. Still, being of an historical turn of mind, I first dug through the historical data on new industries in transport and communications looking for common patterns of growth. They quickly appeared. Rapid early growth rates followed by comparatively sudden maturity occurred in every case. Growth was fastest when the cost dropped while the technical quality continued to improve. When those twin conditions were no longer present, the industry matured. Or the technical quality might still improve, but if the price went up much it cancelled the impetus for new growth. Moreover, it did not seem to matter if the economy was in a recession or not. On this point, the airlines held the biggest surprise of all. During the economic collapse of 1929-33, airline traffic shot up by a startling 500 percent.

In 1960, Keynesian economic theory dominated both government and academia. It focused on aggregate income as the main driver of demand. That income had fallen by nearly half in 1929–33. Thus, the sharp rise in air travel during that same period was a major anomaly that Keynesian theory could not explain. Checking with my economics professors at the University of Washington, hoping for enlightenment, I discovered they were equally puzzled. They had no explanation.

Why did the robust growth of airline traffic throughout the thirties suddenly pause in 1937, the most prosperous year yet since the Great Crash? In 1937, aggregate income had risen nicely. Yet airline traffic growth had stalled. Another puzzle appeared in the data for the next year, 1938, when a sharp recession hit. Most economists then believed it was caused by cutbacks in government spending that had cut aggregate income. The puzzle was that airline traffic resumed its dynamic growth in that same year, 1938.

The latest Keynesian theory could not explain it. However, economist Joseph Schumpeter held that the economic growth that modern capitalism had triggered was a process of "creative destruction" driven mainly by innovations in the hands of the entrepreneurs who created the new industries that provided growth.

In 1933, innovations were coming fast and furious in the aviation industry despite the economic depression. In fact, 1933 was a prosperous year for Boeing because it produced its new commercial offering, the 247, often described as the first modern airliner. Competition drove the airlines to buy better planes because without them, they would lose business. Better airplanes provided airline passengers with a measure of "ecological release" from the prior higher cost and rougher rides of the older aircraft. Those constraints thus relaxed, the airlines saw new opportunities to gain more traffic. The 247 flew faster and offered greater comfort and at lower cost than had the previous airliners such as the rickety Ford Tri-Motor. Passengers responded, despite the depression. Still, the 247 was merely a first step and while much better than the planes it displaced, it hardly provided an unlimited potential for growth. In fact it had exhausted its potential by 1937. Meanwhile, Douglas had been busy with its own modern design called the DC-3, and this plane is usually described as the first commercially viable airliner. It entered service in 1938, flying higher and faster at lower cost, and hence made money even at lower fares. Despite the recession, passengers flocked aboard the DC-3. The 247 was promptly obsolete and sales went to zero. Boeing turned to military contracts including the B-17 Flying Fortress, the plane that made the company famous.

The FROCA process was clearly at work. Innovations changed the culture of airlines the way mutations in DNA changed bio-anatomy, namely by providing better fitness within an ecosystem, and then causing the ecosystem itself to evolve. Ongoing innovations created a series of frontiers, and each one provided ecological release and new opportunities. These were exploited as fast as possible and once exploited, stopped. A crash (airline traffic pause in this case) followed at once. Before the crash did any lasting damage, however, new innovations opened a new frontier once again. New airplanes appeared that flew faster and more comfortably yet, and at a lower cost so the FROCA process repeated itself. While aggregate income clearly played a role in demand, it was not driving the growth of airline traffic. Innovation was.

This innovation analysis explained events right up to the time I was doing this study. Ten years after the introduction of the DC-3 that had spurred rapid growth, traffic again stalled in 1947. Again this proved puzzling. Growth had slowed somewhat during World War II as the Army Air Corps had taken over much of the equipment for military purposes and the airlines' capacity was artificially constrained. Shortly after those planes were replaced, traffic surged for a couple of years. However, it again stalled out, and in the midst of a post-war reconversion boom, again to the surprise of many economists and all the airlines. In 1949, again to the surprise of economists, traffic resumed its dynamic growth just as the 1949 recession hit. In 1948, however, yet a new generation of commercial airlines was delivered. These new planes had four engines, pressurization that allowed them to fly higher and faster and over-fly most bad weather, and hence had much smoother rides. Not only were they faster, (300 mph versus 175 mph) they also had a much longer range. For the first time the more advanced versions of these airplanes (DC–6s, B-377s, and Constellations) allowed a passenger to fly nonstop across the nation and to Hawaii. They could also fly to Europe and the Far East with some intermediate stops. Flying faster over longer ranges reduced trip time and fatigue. People flocked aboard. They continued to do so during the recession of 1954 that went unnoticed by the airlines. Then after ten years, suddenly and once again by surprise, traffic growth stopped. Most shocking of all, it stopped just as the jets had started to come into service.

During this post-war period, most airline economists based their forecasts on Keynesian theory. They did so using statistical association rather than logic. They calculated (more or less accurately) that airline traffic grew at three times the rate of aggregate income, a ratio that had held roughly true from 1949 to about 1959 and concluded that the rise in income had caused the much greater rise in the traffic growth rate. Since economists had no explanation for any of the pauses during boom times nor of the resumed growth during recessions they ignored these anomalies.

Having researched and observed this history, early in 1961, I published my analysis called "The Anatomy of Growth." It laid out the role innovation played in the growth of the airline industry and did so in comparison with the similar experience of several other growth industries. I cited Schumpeter as my reasoning for rejecting Keynes. I noted that Schumpeter focused on growth while Keynes was concerned about economic depression. Schumpeter's theory was consistent with these facts and Keynes's General Theory most definitely was not.

My analysis was well received and so I was asked to make a forecast based upon that analysis and I did. At once it proved an effective counter against the apparent sound logic of Hitchings's analysis.

If I gave that forecast once, I gave it more than two hundred times, sometimes twice or three times a day to various groups. I had in effect made my analytical bones at Boeing. As it turned out, traffic did indeed begin to respond in 1961, and sales of Boeing's jets turned up sharply. My forecast had said that the current pause was temporary; it had happened before, and growth would resume as before given the nature of the jet. It flew far faster, over far longer ranges, at much higher altitudes and thus more comfortably, and at much lower costs. Then why had not traffic responded at once? Because, I argued, the airlines had added a 15 percent surcharge on the price of their jets. Moreover, during the transition from piston planes to the jet, the airlines had been forced to disrupt passenger schedules. That had caused consumer confusion. A joke at the time went: "Breakfast in Los Angeles, lunch in New York, dinner in London, and baggage in Buenos Aires." My forecast also contained a cautionary note that the airline industry might expect another traffic pause about ten years after the jets had entered service. In 1964, my innovation forecast was reissued without change.

In 1963, I completed my dissertation on Japan and received my doctorate, and in April 1964, I received an offer to become a consultant overseas at a much higher salary. I accepted the offer and left Boeing in June 1964. In December 1966, however, Boeing made me an offer to return at my now higher salary and so I did in January 1967. Almost at once I began to learn more about the FROCA process first hand. Before, I had focused on technological frontiers. Now I was about to learn what happens leading up to a crash. At Boeing, I discovered that my cautionary warning that traffic would pause once the jets had exploited their market potential had been quietly set aside. As an explanation I was told that traffic was growing even faster than I had predicted. Indeed it was. Moreover, the government had also forecast a pause-free future in conjunction with its funding of the Supersonic Transport Program, Dulles Airport, and other projects. Furthermore, the airlines themselves had committed heavily to Boeing's new jumbo jet, the 747. All such programs depended on high growth rates continuing for at least another six or seven years, and in fact orders had poured in for the 727 and 737. Boeing's production of new jets had risen from about nine per month to nearly thirty. Douglas had launched its DC-10 jumbo and Lockheed its Trident or an L–1011 jumbo. The euphoria was unstoppable, or so it seemed. The bottom line, I was told, was that to plan for a pause was also a plan to exit the business.

As it turned out, the traffic pause came right on schedule, beginning in mid 1969. The government had to cancel the SST program. The airlines quit ordering new jets and canceled earlier orders. Boeing's Seattle area employment had reached about one hundred five thousand in early 1969. By early 1971, it had collapsed to thirty-eight thousand. Someone commissioned a billboard sign that read, "Will the last person to leave Seattle please turn out the lights?"

My first reaction to all this was outrage. Later, as I began processing these events, I came to see them in terms of evolution. It almost had to happen the way it did. To have planned for that pause would have been to forgo most of those frontier opportunities. Put it in the terms of FROCA (terms that date only from 2001), *pioneers must exploit opportunities on new frontiers as fast as they can if they hope to survive at the end.* These opportunities almost always trigger a race in any competitive environment. Had Boeing planned for the growth pause, the company might not have survived the crash. All those orders, however much in excess, would have gone to Boeing's competitors, and those competing jets would have dominated the post-crash fleets.

Instead, Boeing received the majority of orders by continually raising production rates to accommodate the (un-needed) orders. That was enough for Boeing to survive the crash purely on overseas orders plus spare parts. Later, as the 747 and other jumbos entered service and fares were cut, growth resumed at a more modest rate. In the end, Boeing survived the crisis where others did not and was thus positioned to develop more new jets with far better electronics. That opened a new surge of traffic growth. Production rates once again climbed as growth rates worldwide soared and new markets opened up. But FROCA was still a process in motion, and yet another crash would come. This time, it came as a strong backlash to rapid globalization symbolized by the September 11, 2001, destruction of the World Trade Center and damage to the Pentagon.

The terrorists' weapons of mass destruction were Boeing jets used as flying bombs, highjacked by suicide-terrorists who were financed by oil money. Air travel traffic promptly—and for the first time—plunged worldwide, caused not by economics but by personal fear.

My experience at Boeing, upon reflection and in light of so many similar examples, convinced me of two things. First, given human nature and indeed the nature of life itself, any new opportunity will induce competition to exploit it. Second, such competition will inevitably lead to its overexploitation followed by a crash.

19

AMERICA AS SOLE SUPERPOWER

Technology enabled America to break away from the Eastern Seaboard, cross over the Appalachian Mountains, and then colonize the Midwest and the rest of the nation. These geographical frontiers were closed by 1890. Still, the technology kept emerging to open new frontiers, turning the United States into the largest industrial power in the world by 1900 and by 1930 the most prosperous large nation. By 1991, these ongoing waves of ever newer technology had propelled America into position as the world's sole superpower.

Europe, as discussed, gave birth to modern technology as an ongoing process. We have also discussed why the Bill of Rights helped shift the center of innovation toward the United States by keeping open the door to new frontiers of technology. Furthermore, focused immigration that later became known as the "brain drain" created an "ecosystem" that systematically selected for immigrants who had a greater than average capacity for life on an open frontier—a life that forced one to leave behind many relationships and traditions. This trend had unintended side effects, but it nonetheless helped to reinforce America's lead in innovation. For instance, it selected for people more attracted by the pull of the Lexus than that of the Olive Tree, to use Friedman's metaphor. These immigrants brought with them a capacity for change and innovation uninhibited by a sense that they were disrupting a sacred status quo since they had left extended families, clans, and tribes behind. Moreover, if some of these immigrants discovered that America's open frontier life was unsuitable to them, they were free to return home. In fact, about one third of all immigrants to the United States did return home. The result is that America appealed to select immigrants more likely to innovate on one hand and to be less disturbed by social disruptions caused by the innovations of others. Europeans became more tradition-bound over time because of its gradually shrinking pool of potential innovators and entrepreneurs. Many of America's greatest innovators, entrepreneurs, and scientists were and still are immigrants. Athough most immigrants were initially European, most recently China, India, Korea, Mexico, and other countries have furnished many of America's immigrant entrepreneurs and innovators.

A combination of factors thus combined to promote technology in America. During the twentieth century, America was to use its growing power of technology to help defeat one rival form of government after another. Many historians have noted that World Wars I and II were actually separate stages of the same war, more or less between the same rival powers. It is also fair to say that the Cold War, Desert Storm, and the War on Terrorism in Afghanistan and now Iraq continue the same contest. That contest is one to determine how the high-tech world is to be governed. During World War I, President Wilson said it first. The goal was to make the (modern) world safe for democracy. It is equally true to say these wars, by promoting democracy and the free market system, along with human rights (as defined by Euro-Americans) also have the effect, intended or otherwise, of making the world safe for *advancing* technology.

Jared Diamond points out that since the dawn of civilization, high tech beats low tech every time.[1] If in an oversimplified way, that explains why America has won every war it has fought since the Civil War. America by no means won every battle in those wars. In the Cold War, the United States lost the "Battle" of Vietnam, and fought to a draw in the "Battle" of Korea. But in both of those cases the United States avoided using its deadliest technology to keep from alienating its more cautious and far more war weary allies in the Atlantic Alliance.

World War I began by the accident of a wrong left turn of a limousine that put the Arch Duke Ferdinand in the path of an assassin's bullet. That assassination quickly cascaded into a holocaust.[2] This incident is often used as a good example of chaos theory in action, in that a small change in initial conditions cascaded through positive feedback into a catastrophic outcome. While many things set the stage, this assassination did in fact trigger the war, and had it not happened as it did, the war would not have begun then and perhaps not even later. This point is often noted but is sometimes contested by "determinists" who insist the arms build-up made the war "inevitable." As the Cold War's end showed, it was not inevitable.

World War I did not at first involve the United States and when America did enter the war, its weapons technology was no better than that of either its allies or adversaries. Indeed, the technological power of America at that time lay in the industrial sector. From a military point of view, the United States did have a new state-of-the-art navy that helped Britain's Royal Navy subdue the German U-Boat menace. America could thus dispatch about two million fresh troops to France with no losses at sea at all. Those troops were put into action to help stop Germany's last major offensive of March 1918, which finally failed in August. By October, an exhausted Germany sued for an armistice that went into effect November 11, 1918.

The world's three major democracies—Great Britain, France, and America—had subdued Europe's ruling monarchies. Germany, Austria-Hungary, Russia, and that of the Ottoman Turks all collapsed. Still, the world was not "safe for democracy." Russia became a communist despotism; Germany and Italy became fascist despotisms. In Asia, Japan too had become a variant fascist despotism. In effect, World War II pitted the democracies against the fascists. True, Adolf Hitler forced Russia to ally itself with the democratic and capitalist nations that communists despised. Having first made Joseph Stalin an ally in a joint invasion of Poland in September 1939, Hitler then invaded the USSR in June 1941; so Russia was suddenly willing to cooperate in a common cause. The United States entered the war only after being attacked by Japan at Pearl Harbor and even then Hitler had to declare war first. It is also true, however, the United States and Germany began an undeclared war in the Atlantic beginning in August 1941 when President Roosevelt ordered the U. S. Navy to sink German submarines on sight.

Between the two world wars, however, the United States had made much progress in weapons technology and other aspects of warfare. Its latest naval vessels were among the world's best, especially in aircraft carriers. Moreover, America also perfected radar for its ships, innovated sophisticated techniques for refueling ships at sea, and had developed a wide variety of military aircraft. Its heavy bombers were the best in the world, even at the outset of the war. The fighter planes, laggards at first, were the best or nearly so by 1944. Additionally, America could build far more of them than could the Axis powers. However, America's ground forces, except for the Marines, were far from the best. While the army was well-equipped with motor vehicles, small arms, and artillery, its tanks were inferior to German or Russian tanks. Still, by 1945, the American army was very powerful and had finally caught up in tanks while having an overwhelming air superiority. But the atomic bomb was America's supreme weapon, which brought the war in the Pacific to an abrupt end in August 1945 when Japan surrendered unconditionally.

In short, by the end of World War II America had become a nation powerful in technology and perhaps the most powerful. Still, the British and the Russians could also boast of much. The Germans, moreover, developed long-range ballistic rockets (thanks to an American innovator, Charles Goddard), cruise missiles of sorts, and jet fighters—all too late for dominance in the war.

While fascism had been crushed by the Allies' military victories in World War II, the communist Soviet Union had emerged as a superpower in its own right. The USSR possessed by far the most powerful military force on the Eurasian land mass with millions of troops and tens of thousands of tanks, artillery pieces, and combat aircraft. It quickly built its own atomic bombs and was the first nation to use Germany's rocket technology to launch a satellite successfully into space.

As recently as 1960, economic observers could not determine if free-enterprise capitalism would prevail over Soviet-style central planning. Many Americans believed that central planning was more efficient because it eliminated "wasteful competition and duplication." Why have five or six major car manufacturers when one could supply all the cars needed? What was the point of eleven major trunk airlines when one government airline could do the job? All this duplication produced wasteful competition. Competition prompted the "squandering" of scarce resources with advertising that often promoted goods that "people did not need, and could ill afford," all in the "crass" bourgeois pursuit of profits.

These arguments in favor of central planning were widely believed the world around. They appealed to the intelligentsia everywhere, including in the United States. Many American academicians considered Marxism to be a superior economic philosophy that offered the only practical way to ensure distributive justice and that disdained crass bourgeois competitive values that put the quest for profit ahead of the more refined pursuits of "high culture"—in the arts and literature, for example. With some differences, to be sure, academicians continue to promote a version of the Platonic cultural code of southern plantation owners who took pride in promoting the gracious refinements of civilization over its much messier productive side that was best left to lesser souls.

During the late sixties when many young Americans raised the red flag in protest over the Vietnam War, typical Marxists would not concede that capitalism was the more productively efficient system. They attributed the apparent lag of the standard of living in communist nations to a late start on industrialization and to recovery from the ravages of World War II. As for technology, they could point to spaceship Sputnik, the speed with which the Soviets developed the atomic bomb and ballistic missiles, and even the AK-47—a favorite weapon in protest movements. What they overlooked was that all these Soviet accomplishments drew on the innovations made by others, not by the Soviets.

The technological superiority of the West, however, gradually, became evident as the Cold War wore on. That "war" is usually thought to have started with the Berlin Blockade and airlift in 1948. At once American and British logistical air power became evident. For many months Berlin was supplied entirely by air, including coal for heating in the chill winter months. It was seen as such a miracle of logistics that the Soviets finally backed down without concessions from the United States.

The hot side of the Cold War was fought outside either nation. The first big test came in the Korean conflict, when for the first time American jets fought Russian jets. Although the MiGs were effective fighter airplanes and had advantages over the American F-86 Saberjet, the win/loss ratio was heavily in America's favor. A repeat performance took place in Vietnam during the sixties and seventies with later versions of both nations' jets. The outcome was similar. Later, the conflict in Palestine between the Arabs and Israelis put a variety of western and Soviet weapons systems toe to toe. The results were the same, and the West's technical superiority became more pronounced as time passed. Better training accounted for much of the western success in these conflicts but not all. By the mid eighties, America had a clear advantage in almost every category of weapons systems.

Advancements in electronics made the crucial difference. Here the Soviets developed innovations based on captured German technology. However, British and American electronic technology already had been superior in World War II. The Germans and Japanese were at a disadvantage in radar and elsewhere. The electronic computer was an Anglo-American invention during the war and after the war progressed to the mainframe computer based on computer chips. Technological superiority began with the American invention of the transistor that replaced the vacuum tube. The transistor was lighter, less expensive, and more reliable than the heavier, costlier, and less reliable vacuum tube. The quality of radio, television, radar, and computers all rose sharply, and costs came down. As transistors got smaller and cheaper the cost per unit continued to plunge according to Moore's Law, named for Gordon Moore of Intel.

Thanks to electronic computers, America began to pull ahead in the "space race." The Soviets launched Sputnik in 1957 to orbit the Earth, and America was far behind. But twelve years later, in 1969, American astronauts had landed on the moon. The Soviets could not match that feat, but not for lack of rocket science but for lack of electronic science. They could not compute accurate trajectories without powerful computers. Moreover, America launched the first communications satellites, also made practical by transistors and chips, which also drove down the cost of long-distance telephone communication.

By the 1980s, thousands of transistors could fit on a small computer chip. The hand-held electronic calculators, personal computers, cell phones, and an array of new electronic gadgets emerged from the market, one after another. So did precision-guided bombs and cruise missiles. Information technology advanced so rapidly in America that by 1990, more workers processed information than they did raw materials. A computer terminal in a factory or office quickly became an almost ubiquitous American workstation.

These knowledge workers had displaced Marx's beloved proletariat. Machines took over the heavy lifting and computers took over the heavy number crunching. The Soviets came to realize that for all its theoretical advantages, central planning in a command economy did not work. Shortages and continual mismatches between supply and demand became a routine fact of centrally planned life. Collective farms had proved a dismal failure, as Russia, once a great exporter of grain, was forced to import grain from the United States and Canada. The Soviet workforce was sluggish and badly motivated. *"They pretend to pay us and we pretend to work,"* was a communist in-joke. Practically nothing the Soviets made aside from the AK-47 could compete on the world market.

A Boeing 767 flying from Warsaw to New York used less fuel than the equivalent Soviet model flying from Warsaw to London. That example illustrates an essential difference in the two systems. Having substituted administrative prices for market prices, Soviet jet designers had no motive to design fuel-efficient jets. In any event, Aeroflot flew the planes they were given. In the name of capitalism, western airline companies forced manufacturers to design and build better airplanes to fit their specifications. To a Marxist such competition produced wasteful duplication of effort. In reality, it served to elicit the best efforts every bit as much in a competitive business as it would in a competitive team sport. Competition sometimes may be wasteful, but its absence greatly reduces the motive to perform well. Most people need a strong motive to exert themselves to the best of their abilities.

America's comprehensive technological prowess in the military arena became obvious to most people during Desert Storm in January 1991. Memories of Vietnam still remained of drug-besotted draftees, faked body counts to please the higher-ups, and very low morale. By the late 1980s the U.S. military had reinvented itself. By then it was an all-volunteer force with much upgraded standards of enlistment and vastly improved training and was re-equipped with many new and more sophisticated weapons such as the Bradley Fighting Vehicle and Abrams main battle tank. The Air Force and Navy had developed a suite of precision-guided weapons, stealth airplanes, and much more. During Desert Storm, the results were almost shocking. After a few weeks of air war the actual land battle was over in just one hundred hours. No ships were lost, only a few airplanes were shot down, and almost no tanks were destroyed. By most calculations the U. S. military suffered a lower percentage of deaths in combat during Desert Storm than if the same troops had stayed home and died drunk in auto accidents.

Desert Storm ended in February 1991. In August of that year the Soviet Union was finished, a failed experiment in central planning. It disestablished itself and broke up into several independent republics, with Russia the largest. East and West Germany reunited. The Cold War was over; a nuclear holocaust had been avoided. America stood triumphant, the world's only superpower. All competing despotisms had seemingly either been defeated (absolute monarchy in World War I and Fascism in World War II) or had quit the contest as did China and Russia, if in different ways. Only Cuba and North Korea remained to carry forward communist central planning.

By the time of its collapse, the Soviet Union had come to see the pivotal weakness of a centrally planned command economy for itself. It simply could not innovate and in effect was designed to preclude the possibility of innovation. All information was tightly controlled. All major decisions were made at the top. The concept of entrepreneurship was bureaucratically impractical. In principle, a worker on a project could submit an idea; but to receive final approval the idea required approval from as many as thirty-five levels of interim management and rarely passed level three or four. The KGB was perhaps the first major Soviet agency to become aware of this fact, thanks to forty-five years of highly effective spying. In fact, the KGB saw the truth more quickly than did western economists in thrall to macroeconomics. Aggregate income, aggregate goods supply, aggregate money supply, aggregate levels of investment both in research and development and in brick and mortar were the focus of their attention in macro-economist's statistical macro models. This aggregate focus systematically lost sight of what happened at the grass roots margins of the economy where most innovations took place and where nearly all entrepreneurs lived to energize bottom-up economics in the spirit of Joseph Schumpeter. Keynesians and monetarist macro models had a top-down focus that lost sight of the bottom up action.

What KGB spies had uncovered over and over again was the vital role played by the famous "garage start up" in the American economy. Individuals created the innovation, and just as important, brought it to market via entrepreneurship. This was particularly true in electronics. Hewlett Packard, Intel, Apple, Microsoft, Netscape, Google, Dell Computer, AOL, and many others began in a private garage or a near equivalent. Some of these entrepreneurs were teenagers or young adults in their early twenties (Bill Gates, Steve Jobs). Others, such as Gordon Moore and Andy Grove of Intel, were refugees of corporate bureaucracy. As did thousands of other creative people, they discovered hierarchical corporate conservatism tended to quash their best ideas or did nothing with them. So they started their own firms and many became hugely successful, as the KGB duly noted.

The Bill of Rights provides a protected space for these innovators and entrepreneurs from corporate conservatism. Large business corporations, governments, and NGOs (non-governmental organizations) grow mainly by expanding their territories. Sometimes they absorb innovative firms using the latest technology. At other times they innovate, although not always to their own advantage. For example, Bell Labs invented the transistor and Xerox invented the mouse. But it was not AT&T that promoted the transistor and it was not Xerox that exploited the mouse. Xerox practically gave the mouse to entrepreneur Steve Jobs of Apple who used the mouse to create the Macintosh computer platform.

As a rule, groups with power aim mainly to protect their power. Innovations, even their own, pose a potential risk by disrupting internal power relationships or creating vulnerability to competitors who copy it. Many scientists who argued against making the first atomic bomb focused on that latter theme. As stated earlier, almost as soon as they formed, despotisms in various areas of the world constrained innovations in different ways. If they expanded geographically, it was by conquest, an early version of a corporate takeover. Conquest preserved order and stability by subduing competitors or subduing lawless border dwellers who threatened them. Conquest continues today in business settings and motivates most corporate mergers and takeovers. Such mergers, however, can sometimes weaken rather than strengthen the merged firm.

This risk is routinely overlooked. For example, when Boeing merged with Douglas, many bitter Boeing employees complained that Douglas took over Boeing with Boeing's money and then went on to destroy a history of good labor relations. Something similar took place with Time Warner/AOL, Daimler-Benz, and Chrysler. Unfortunately, those who analyze the financial results of a proposed merger often ignore the issue of business culture compatibility—a measure of possible success or failure that is not calculated on paper. Not until after the merger does the problem become evident and by then it may be too late.

Almost all our large American corporations began as small start-ups based on an innovation that an entrepreneur thought worth attempting to exploit. The small company operates on the frontier part of FROCA. If it becomes big and survives any intermediate crashes, it enters the adaptation stage. Once there, it aims to achieve and maintain stability and will subdue threats to such stasis if possible. The larger company may continue to grow with a rising tide of the general economy. It may also aim at continuous incremental improvement and development of existing technology. But it is rare indeed for a large corporation to invent and then actually use new technology that represents a paradigm shift. The overwhelming cultural pressure in any well established entity is to suppress potential paradigm shifts, which is why, even in a free-market, free-enterprise system, significant innovations develop on the periphery. That being the case, the Bill of Rights is needed to prevent the culture from suppressing innovative and entrepreneurial fringe dwellers.

America's Cold War and Desert Storm victories did not bring liberal democracy to the world. They did not put an end to the history of ideological conflict as Francis Fukuyama put it. Rather, they leave open the question of who will rule the world and how? We must now ask this question knowing that the very technology that lifted America to its present supremacy and helped limit monarchy, fascism, and communism now permits a worldwide communication and information exchange. American technology helped create the electronic global village, which in turn put an end to the geographical separation that long allowed different cultures with quite different values to live apart without knowledge of larger world issues. Our technology has eliminated that separation. Olive tree cultures now view Lexus cultures as jabbing the fingers of technology in the eyes of their sacred olive tree values and traditions.

In sum, we seem to have entered a world of high-tech anarchy in our new electronic global village. It brings a certain sense of déjà vu. Today's anarchy in some ways resembles that of six thousand years ago in agricultural villages. Recall that once their population had grown because of agro-technology, they began to murder each other in family feuds. The old egalitarianism could not adjust to a new world order of large populations of many strangers living in close quarters without a government to impose law and order. Thus, the people affected by that anarchy took progressive steps toward despotism to restore law and order, a despotism that nearly stopped the innovations that brought forth the new technology. In effect, our current world order seems to be recreating this same problem but in the context of a much higher level of technology.

How will it all work out? Are we like the Norse of 1400 Greenland who refused to eat abundant fish in favor of raising the cattle that destroyed their local ecosystem and caused them all to die?[3] Are the world's olive tree cultures engaged in a last ditch stand, like the American South in the Civil War, fighting a losing battle to protect traditions, values, and a way of life that cannot survive high-tech democracy?

What are our choices, and how will we make them?

PART V

TECHNOLOGY AND THE PROSPECTS FOR A META-CRASH

20

GLOBALIZATION AND ITS DISCONTENTS

Globalization took its first tentative steps when Europeans such as Vasco da Gama, Columbus, and Magellan used more sophisticated technology to explore the world. The Europeans later established trading posts, then colonies, and finally imposed their technology on the areas they conquered. None of that was new. First came the traders, and then came the soldiers. What was new was that no empire was able to expand on a global scale before Europe invented its new technology that broke out of the ancient mold. Every earlier empire, even Rome's, had exhausted its resources well before coming close to being global.

This first stage of globalization, 1500–1800, ended after the industrial revolution began. The imperial powers—starting with Britain—then had to find markets for their industrial output—starting with textiles—to pay for the needed raw materials such as cotton that they did not themselves produce. This exchange enabled the colonies to buy much less expensive goods than those they had been producing by hand. But those cheap goods often devastated the cottage industries of those colonies. For example, resulting resentment in part spelled the doom of imperial rule in India. Mahatma Gandhi used this issue with great effect in his campaign to persuade the British to leave India and thus allow self-rule.

Nevertheless, imperialism laid the groundwork for some colonies such as Singapore, Hong Kong, Taiwan, and South Korea to become modern high-tech nations in their own right.

In general, colonial powers upgraded the local infrastructure of their colonies by building telegraph and railroad networks, improving the roads, ports, and often public sanitation. They created a more modern administrative apparatus of government. Some colonies used their imperial legacy as a springboard into the modern high-tech world, but others did not or, perhaps, could not.

This second stage of globalization lasted slightly more than one hundred years and peaked in 1914, ending abruptly at the outset of World War I. Stage Two generated a robust level of world trade and used gold as an international medium of exchange. Progress seemed to be on the march the world around, thanks, in the view of the Europeans, to their colonial empires that brought modern western enlightenment to the rest of the world.

The United States engaged in globalization as it peaked in its second stage. As a result of its war with Spain in 1898, the United States took control over the Philippine Islands, Puerto Rico, and Guam. From the first, the United States planned to withdraw from the Philippines and did so more or less on schedule. The well-established empires included those of Britain (on whose flag the sun never set), France, Germany, Austria–Hungary, Italy, Russia, Holland, and Belgium. Toward the end, Japan attempted globalization when it seized Formosa in 1895 and Korea in 1910. The European powers were mostly industrialized and Japan soon joined them. Steamship lines and telegraph networks were beginning to integrate communications and transportation around the world. Economists clearly saw that the imperial–industrial world was increasingly interdependent. Many economists hoped such interdependence would render further war obsolete by making it too costly. High-powered, long-range artillery, rapid-fire field guns, and machine guns also seemed to make war much too deadly.

World War I dashed that hope as it also put an end to Stage Two of globalization and stalled it until the Cold War ended. After World War I, America's Great Depression struck, followed by World War II, and then by the Cold War. These events inhibited globalization on the political level, although the underlying technology continued to move ahead at a rapid pace.

After World War II, all the great European empires (except the Soviet Union's) began to detach from imperialism. The British withdrew from India in 1947. Although Britain provided help to some of its former colonies that became independent, France, Holland, Belgium, and Portugal were ejected by force for the most part. Italy and Japan lost their colonies as a result of World War II. By 1960 the Golden Age of Global Imperialism had passed. Out of the old colonies arose most of the Third World nations, and their independence, interacting with the Cold War, slowed globalization to a crawl.

These new nations varied in how much modern technology they took from First World countries. The new Asian states took quite a bit, seemed to thrive on it, and still managed to retain many of their old traditions. States friendly to capitalism such as Singapore, Hong Kong, Taiwan, South Korea, and Japan enjoyed remarkable progress. However, nations that embraced Marxist ideology stagnated, and they remain stagnant. Nations that began independence with a Marxist bias, and later let it go, then made progress. India and China are good examples.

While globalization was on hold, those nations that embraced capitalism with its technology were able to modernize at a pace of their own choosing. Thus, they were able to accommodate many of their core values and traditions to the modern world without wrenching cultural trauma. Little of today's backlash against resurgent globalization that began when the Cold War ended comes from these nations.

Today's backlash centers in the Arab nations and includes some others that embrace Islam. This puzzles many people since Islam was at one time the most advanced empire on earth. It stretched from the Atlantic Ocean to the Indus River, encompassing all of North Africa, Iberia, Iraq, Arabia, Persia, and many islands in the Mediterranean Sea, together with Afghanistan and part of central Asia. It had the world's most advanced science and math at the time (roughly 800 to 1200 CE) and an architecture still much admired and copied. Our numbering system and algebra came to us from the Arabs. Much of the learning the West recovered from the Greco–Romans was possible because the Arabs had preserved the works of Plato and many other classical writers.

Islam was in many ways a more tolerant religion than early Christianity. Moreover, by comparison, Europe was so backward as to be almost beneath Islamic contempt. The Arabs hardly paid Europe any attention. Islam's concept of Jihad was then more a program for reform than today's fundamentalist concept that advocates violence and terrorism. Little of this violence and terrorism finds support in the Koran, according to most Islamic scholars, such as Armstrong and Lewis, earlier.

So how and why could Europe seem to come from nowhere to overtake and later overwhelm Islam, once the largest and possibly most powerful empire on earth? (I have addressed the question of Europe's recovery, but not the question of why Islam failed to keep pace with Europe's surge of new technology.) The Ottoman Turks did adopt European artillery and other firearms. But at the same time they pointedly refused to adopt the printing press for nearly three hundred years, according to both Armstrong and Lewis.

Both scholars addressed this question in their recent books.[1] One reason was that the Koran was written for a nomadic tribal culture of herders that retained much prehistoric egalitarianism. Very little of Arabia was suited for farming and there were no large agricultural surpluses. The Arabs also had a strong cultural distaste for strong monarchs, besides having no large surplus to tax. Additionally, nomads do not lend themselves to central government control. As Armstrong stresses, resource scarcity encouraged the tribes to engage in almost continuous tribal warfare, mainly by raiding each other.[2] Raids, in fact, were a common practice in tribal communities throughout the world. Still they were disruptive to the two key cities of early Islam—Mecca and Medina. Both cities were trading centers, and the local terrain supported some agriculture.

One of Muhammad's hopes as his Koran was written was that Allah would provide the tribes with a common focal point for a community of faith that would end disruptive tribal warfare. Indeed Muhammad hoped to join the Christians and Jews into that same community of faith and as a result to end violence, not to promote it. The three faiths did not merge and instead each followed its own holy scripture as God's final Word. Muhammad did manage to create a community of faith among the nomadic Arab tribes, however, and he reigned as both spiritual and temporal leader. He did not style himself as a monarch.

Yet Muhammad's very success in creating such a community created an unintended crisis. Tribal survival had come to depend on raids for necessary supplies. Since the faith placed strong inhibitions on violent raids of their Islamic community, the tribes were pressed to find a substitute. They began raiding outsiders and were so successful they soon broke out of Arabia itself. When they discovered the opposition was weak, after the fall of Rome, they found great early success in that region. The Arabs pushed outward, exploiting an ecological release from Rome's defunct legions. They pushed right to the Mediterranean and quickly took Damascus and Jerusalem. The raiders sent booty back to Medina and were able to finance a more or less regular army. With that supply, the Arabs pushed on to Egypt in the west, and into Iraq and later Persia heading east. The Byzantines and Persians had weakened each other through war so that the Arabs moved in and conquered Persia.

The whole North African coast was almost defenseless and the Arabs pushed all the way to Gibraltar. At first they had no intent to convert the locals in the regions they had occupied. The locals—mostly Christian, Jews, and a few pagans—were allowed to continue to practice their religions. Indeed, the Arab armies kept to themselves. The Arabs continued until they had crossed over the Pyrenees into France where they were defeated in the battle at Tours-Poitiers in 732 CE, scarcely one hundred years after their advance began. The defeat ended Arabian attempts to eliminate a strategic threat north of the Andalusi border. More significantly, the battle's loss stopped Muslims of Islamic political and cultural beliefs from spreading further into formerly Christian–Roman lands and allowed for regrouping of Christian dominance over the following seven centuries.

Muslim religion seemed to rule out a traditional emperor. Moreover, the Koran, and not a ruler, was thought to be *the* divinely ordained source of all law, the Shariah. Koranic law should guide behavior in all aspects of life in the political as well as the spiritual arena. Christianity, by comparison, allowed for the separation of the state from religious affairs. That view was based on Jesus Christ's words in the Gospels that one should "Render unto Caesar the things that are Caesar's and unto God the things that are God's" (Mark 12:17).

The Shariah might have worked well in a community of nomadic tribal people living with seventh century technology and who had no king and did not want one. It did not prove a workable proposition for running a large empire acquired over a short amount of time.

The first Islamic rulers were called caliphs, and they soon shifted the caliphate from Medina to Damascus, then to Cairo, and later to Baghdad. From the outset, a schism arose as to which of two family lines should inherit Muhammad's mantle. That division remains to this day, the Sunnis on one hand and the Shiites on the other. Over the centuries, Muslim rulers have had various titles such as caliph, sultan, or pasha. But under any title they were then (and still are now) forced to act like despots and make their own laws or rules apart from the Shariah and despite the Koran. At the same time, the rulers typically became kleptocrats who lived in grand luxury in the fashion of other despots. Booty from foreign conquests supported such luxury. These departures from the Koran invited a backlash from Islamic religious leaders from the beginning.

Still, it never proved possible to rule an empire with the Shariah for lack of a central human authority who could make flexible decisions based on changing circumstances. One section of the early Arab empire, Iberia, soon broke away. Islamic disjunction between political reality and religious revelation may have partly fostered the cultural glory days of Islam. Before various forms of interpretive dogma began to appear, a certain freedom of thought bloomed. However, Arab rule began to weaken, first under the impact of the Mongol invaders from Central Asia. With no dominant religion uniting them, Mongols, then Moguls, and finally Turks all embraced Islam, each group with its own non-Arabic cultural emphasis. These invasions began about 1200 CE. The most successful and long-lasting invaders were the Ottoman Turks. Under Turkish rule, Islam reached its highest political influence. The Turks took Constantinople in 1453 and twice reached the gates of Vienna. The second attack in 1683 proved a disaster. The Turks were badly defeated, and the Islamic tide turned. By 1683, Europe's resurgence gained global attention, even of the Arabs who had long ignored Europe as a region from which they had nothing to gain. More than three hundred years later, Islam still struggles with resentment over western technological superiority and the conflicts inherent in its own underlying secular political and cultural focus.

By comparison, Japan was confronted by Commodore Perry with far superior western technology in 1853, and was faced with a choice of embracing that technology and using it to defend its own sacred traditions or risking colonial dependency. Japan's pragmatic strategy proved highly successful, despite the debacle of World War II. Historically, Islam's seeming unwillingness to accommodate the Koran to the modern world may be part of the reason. Japan's advantage was that its various religious and ethical beliefs such as Zen, Shinto, Taoism, and Confucianism contained little dogma. "The Tao that can be spoken of is not the real Tao." Islam, by comparison, was rich with rigid dogma spoken of as revealed truth, and many reformers responded to the western challenge with that dogma.

True, in the twentieth century some Muslim governments tried to develop along secular, but almost never democratic lines.

Turkey after World War I was the first example, followed by the Baath party in Syria and Iraq. Egypt under President Gamal Nasser after World War II made an attempt to create a secular government. Those secular efforts have been condemned by fundamentalist Islamic religious leaders. In principle, the idea of nationalism as a focal point of loyalty and personal patriotism is deeply anti-Islamic. Islam should be the sole focus of loyalty under the direct rule of Shariah and without some separate set of secular civil rights that protect individuals from the government.[3] According to most Islamic scholars, Shariah is supposed to take care of all our needs. To hold otherwise is to claim it cannot give care which is to suggest the Shariah itself and the Koran from which it comes is also flawed. Thus to suggest that modern circumstances make the Shariah obsolete is to suggest the unthinkable. It is to suggest that the Koran was not, after all, God's final revealed truth to humankind. That in turn is a high heresy that places the whole Islamic project in doubt and can lead to an individual's assassination. The seeming intransigence of Islamic fundamentalists has a certain logic.

The bottom line is there should be no separate states with separate sets of laws that divide the loyalty of the faithful. That secular and nationalistic approach is the way of the hated infidel, in the fundamentalist view. But that same secular and nationalistic way also led to the vast technological superiority of those western infidels. Islam is thus in a psychological bind here that is specific to Islam. Broadly speaking none of the eastern religions has this serious a problem in coming to terms with modernization while at the same time trying to preserve its own traditions. Japan, for example, never felt that the effort to modernize, using the means necessary to modernize, would in and of itself render the "sacred spirit of Japan" invalid. Instead, Japan modernized as the only practical way to preserve its sacred spirit.

Thus, the violent backlash to globalization has arisen mainly in the Middle East, where Islamic, and mainly Arab, fundamentalists have their greatest strength. On this issue, the existence of Israel poses a special dilemma. In a way, Israel is at the center of an intensely bitter family feud. While Islam considers both Christians and Jews to be "people of the book," the Jews are "family" in a sense the Christians are not. Both the Jews and Arabs regard Abraham as their joint founding father. One son, Isaac, founded the Jewish line; the other son, Ishmael, founded the Arab line. Christianity separated from Judaism in part by relaxing the Jewish religious law and accepting new members who did not follow that law. Over time, Christians represented a conglomerate of tribes with numerous genealogical lines. In the early days of Islam, the Jews were treated well. Jewish religious law had much in common with the Shariah because it also prescribed a body of rules for living daily life designed to keep the faithful ever mindful of the presence of God. Some of the rules were the same, for example the prohibition against eating pork. Moreover, adherence to these rules had the effect of separating Jews from Christians after the Jewish diaspora.

Since the beginning of Zionism in the early twentieth century and especially since the founding by the United Nations of the state of Israel after World War II, Israel posed a threat to Islam. Israel began as a western-style democratic nation-state with many secular overtones. As such, it stood as something of a symbol of all the Koran stood against. Additionally, European Jews participated fully in the emergence of modern secular science. At times they dominated it, as with Albert Einstein and Neils Bohr, and many others. Jews had also risen to the top levels of western medicine, finance and banking, and if not in manufacturing, then at least on the marketing side of capitalism. When independence came in 1948, five hundred thousand Jews decisively defeated vastly greater numbers of Arabs. To humiliate an opponent without actually subduing them usually leads to a greater resolve to get even. If anything subsequent Arab defeats at the hands of the Israelis after 1948 led the Arabs to become more intransigent in rejecting the presence of an Israeli state. It was instead the beginnings of Arab terrorism, first against Israel, then against westerners in general, and against the United States—Israel's most faithful ally.

Never mind Palestine as the ancient and sacred Jewish Promised Land, and never mind the German Holocaust. Israel was, in a sense, a preview of post–Cold War globalization in that it posed a threat to the basic sense of Arab Islamic identity. The Jews were "first cousins" who had become apostates within the Semitic family. Thus a Jewish state organized along western lines, using western science and technology, was seen by fundamentalist Arabs to threaten Islam values. Currently, other Islamic peoples tend to support the Arabs on the issue of Israel without the passion of an in-family feud. If Arabs today could accommodate an Israeli state, so could Persians, Turks, Albanians, Berbers, Indonesians, Pakistanis, Bengalis, and those tribes of the Islamic faith in tropical Africa.

Arab fundamentalism is not the only major source of backlash to the recent surge of globalization. To many people around the world, globalization is merely the latest threat a rapacious capitalism poses to the global ecosystem. The most vocal of these environmental activists are Americans and Europeans. Their "green" movements also attract many traditional opponents of capitalism and western imperialism. Environmentalism incorporates several different themes, some inconsistent with each other. Moreover, it contains a deep philosophical contradiction almost as dramatic as that of Islam. For instance, capitalism is accused of severely compromising the world's environment by its huge outpouring of automobiles and other polluting forms of both consumer and capital goods. Because of capitalism and the technology it promotes, humans now consume about one hundred thousand times as much as they did eight thousand years ago and that consumption is growing. Capitalist technology not only supports one thousand times as many people; each of them on average now consumes nearly one hundred times as much.

This record is hardly consistent with the second accusation. Capitalism, its critics complain, creates the unequal resource distribution of a relative handful of the very rich, such as Microsoft's Bill Gates with roughly one hundred billion dollars, while homeless street people subsist on handouts. Such extremes do exist but for many different reasons. Ever since high-tech agriculture made possible the rise of large populations, governments have been structured that enforce large differences in wealth or income from top to bottom so that a small group at the top receive a disproportionate large share of the wealth. No system yet devised has preserved the egalitarianism of the hunter–gatherers for large populations. Yet capitalism has raised living standards so high as to make it possible to eliminate poverty. No other system has yet matched that record.

The complaint about too much consumption makes a valid argument that capitalism threatens the global ecosystem. Huge increases in consumption have indeed entailed huge increases in pollution and at times enormous damage to existing local ecosystems. For example, in China a wall of smog blankets much of the country as a result of rapid coal-fired growth over the past twenty years. It is also true that, as presently organized, capitalism requires continuous growth to avoid a major crash. Recall from Chapter 17 how the automobile industry collapsed in 1930 because demand stalled. Manufacturing output had to drop by 50 percent when the growth in registrations ceased. Replacement demand alone cannot keep the capitalist system fully employed if it has geared up to serve robust growth.

Of course, every new technology must at some point mature. However, if new technologies emerge that open even newer frontiers of growth, a meta-crash is avoided. In this case, self-sustaining technology becomes a rising tide that lifts all the boats it does not sink. The problem is that at some point that same rising tide may flood the global ecosystem.

For example, neurobiologist William H. Calvin and others have suggested that global warming could disrupt the Gulf Stream so as to trigger a new ice age of sorts by cooling Europe's climate down to that of present day Canada and thus reducing Europe's ability to grow food by perhaps 60 percent or more in a mere five or six years.[4] Environmentalists, however, are well aware that it can be off-putting to the public if activists make average affluence the enemy of the global environment. It thus helps to make capitalism, in the hands of a few greedy plutocrats, the villain. Charge them with creating enormous disparities of wealth, thus creating enormous distributive injustice. This attack ignores the 401k plans, other forms of pension plans, and tax deferred savings accounts that over the past fifty years have put the great bulk of common stock into the hands of the general public. If capitalism is really the villain, then to paraphrase Oliver Hazard Perry, "We have met the enemy and they are us."

A related concern that environmentalists face compares to the dilemma Islamic fundamentalism has yet to resolve. Fundamentalists must adopt much of their enemy's technology to fight them. Yet they dare not consider adopting the same system used to create and then sustain that technological lead because to do so would negate the logic of Islam itself. Environmentalists, for their part, discover they must defend the Bill of Rights to ensure their freedoms to speak out. Yet that same Bill of Rights is what sustains the growth of technology under capitalism. It thus accounts for the huge increase in consumption and collateral pollution, and hence poses an ongoing threat to the global ecosystem. Without that Bill of Rights, large corporations would, like all authoritarian governments, and indeed self-organized ecosystems, strive to preserve equilibrium. Innovative new technology, like new immigration, always poses a threat to system equilibrium. Without the Bill of Rights capitalism's inherent conservatism would soon become a friend to the environment. Cartels could form to damp down most competition by fending off, quashing, or absorbing most new rivals. They could set prices at a level of "all the traffic will bear," that would maintain the health of the least efficient. Such prices will mean lower consumption and thus less pollution. What is more, the more fairly the corporations distributed their profits, the more conservative they would become. Technology would no longer become a self-sustaining force. As new technologies matured, the frontiers would close down and stay down. Far fewer innovations would arise to replace them. The four-hundred-year-long period of techno-cultural punctuation would be replaced by the equilibrium stage. Stable ecosystems reject most innovations and mutations alike. The olive trees would take control and once again be safe from brash technological assault. New frontiers might open only as a result of climatic shifts or asteroid hits that were not caused by humans.

Fortunately, the Bill of Rights remains in force, defended by those whose anti-technological cause is a cultural artifact of those same rights. Nevertheless, high-tech globalization is apt to continue to threaten the olive trees of the world and thus engender a backlash. Perhaps the real "new world order," is a resurgent order of non-western tribalism akin to what Samuel P. Huntington suggests.[5] We also face high-tech anarchy amid a possible collapse of the western Alliance.

21

RESURGENT HIGH-TECH TRIBALISM AND A CRASH OF THE WESTERN ALLIANCE

The Western Alliance (also know as the North Atlantic Treaty Organization, NATO) was formed after World War II to protect and unite allied western members against the Soviet Union and its allies in the event of a military attack. Also, U.S. and European countries shared concern over expansion methods and policies undertaken by the USSR. When the Soviet Union crashed in the summer of 1991, however, the main purpose of the Alliance was called into question. Although during the Cold War alliance members set aside many differences with each other in the interests of a common defense, France, for example, resented America's role as the Alliance superpower even before the USSR collapsed and preferred to work for world change through NATO. France harbored the hope that a European Union would provide an effective counter to America—at least in economic and cultural affairs—and put Europe on par with the United States. The collapse of the USSR could have provided France and other European states a sense of ecological release from American dominance, except that the fall of the Soviet Union made the United States the only superpower. According to one poll, the percentage of the French who viewed the United States favorably dropped from 54 percent to 35 percent between 1988 and 1996.

French dissatisfaction over U.S. power has marked French-American relations since the end of World War II when French President Charles de Gaulle held American President Franklin Roosevelt accountable for refusing earlier to recognize the Free

French resistance over the Vichy regime. De Gaulle consequently pursued détente with the Soviet Union, withdrew militarily from NATO, and established an independent French nuclear force, despite the fact that France remained an "ally" of the United States.

Part of the incompatibility between the countries stems from historical tradition. The English language displaced French as the primary international language of diplomacy. That shift troubled the French, because well into the twentieth century, French had been the dominant worldwide language in upper political circles. Further, France believed that a handful of statesmen should make the decisions both for the country and in matters of international policy just as was done in Europe during the nineteenth century with the Congress of Vienna. At the core of current tensions between America and France lay perceptions of one another. America sees France as a once-great power in decline. France sees America as an empire-building country with a need to dominate the world in its own image. Both nations view themselves as worthy of power, a seat of democracy, and culturally and governmentally superior to others.

America saved France from a defeat in World War I and rescued it from a humiliating defeat early in World War II at the hands of Germany. At the time there was much gratitude to be sure, but as that event faded, French resentment grew stronger. If for different reasons, the French shared much of the deep resentment of America felt by Arabs. France identified with the Arabs' pain so to speak. Well before the Cold War ended, France began courting the Arabs and detached itself from support for Israel.[1] After Desert Storm, France began helping Iraq in various ways. It tried to block American efforts via diplomacy to force Iraq to comply with the armistice.

Still, French resentment was simply the most vocal. After the USSR collapsed the Germans began to feel free to express their resentment of American power in various ways as well. In effect, all Europe resented American preponderance. All Europe felt that American high-tech cultural influence was in many ways corrosive and even dangerous. Many Europeans, highly cautious after two devastating world wars, feared America's "cowboy" psychology of the open frontier. Europeans routinely labeled Presidents Reagan and the second Bush as cowboys who shot from the hip. America's dynamic psychology of the open frontier, comfortable with change and even pushing for it, clashed head on with Europe's far more cautious psychology that favors stable equilibrium and thus resists change. The frontier psychology focuses on results. The psychology of stasis focuses mainly on process to preserve order and stability. Frontier values allow for short-cut processes to achieve results. In stasis any result is acceptable if the proper process has been followed.

After early reluctance, America pushed hard to stop the Serbs and their program of ethnic cleansing in the Balkans. Many Americans felt disgust, after watching UN soldiers stand idly by, while Serbs slaughtered groups of their ethnic rivals. The UN troops complained they had no authority to act. When it came time to act, the United States did most of the heavy lifting but for practical purposes suffered no casualties when stopping the Serbs. The same had been true in Desert Storm. Then again in Afghanistan the Americans ousted the Taliban with few casualties. That victory was soon followed by America's quick defeat of Iraq's military, once again with few casualties. In occupation, however, Arab insurgency was quick to assert itself.

America's recent record of military combat success generated more resentment than admiration. In fact, that success sparked much fear. Many nations feared that America under the second President Bush had become a global gunslinger with precision weapons. America possessed smart bombs that could blast away any and all opponents without American casualties. The United States, many nations feared, had appointed itself as police of the world that aggressively looked for altercations at every opportunity to further advance its power.

Outside academia, Americans responded by increasingly seeing Europeans as an "ungrateful lot" of "overly cautious has-beens." True, Europe had been traumatized by two terrible wars, but World War II was seen as a monument to their excessive caution. France or Britain could have stopped Hitler in 1936, according to William L. Shirer because both were militarily stronger than Germany in 1936.[2] But both countries were unwilling to oppose Hitler's plans until Germany had gained strategic advantage and Europe suffered more than twenty million dead in its most terrible war. The allies were only saved from certain defeat by the Americans. After the Cold War began the United States also paid by far the largest share of the cost in defending against the USSR. According to many critics, the Europeans learned little from their own excessive caution as they showed in the Balkans and again with Iraq. Moreover, Europe is no longer the world's predominant culture. The decisive action has shifted to America and Asia and now includes China as a possible emerging superpower.

Meanwhile, in a world of atom bombs, chemical and biological weapons, and even conventional weapons of enormous precision and sophistication, the world has become too unsafe to tolerate terrorist groups such as al Qaeda or Hamas. To tolerate brutal megalomaniacal dictators such as Saddam Hussein or Kim Il Jong is also to court risks. These and other tyrants, such as Idi Amin, have demonstrated by their treatment of rival tribes a taste for tribal genocide or at least brutal mass murder.

Thanks to modern technology today's fragmented world is thus an increasingly dangerous place. Cautious European diplomacy may only escalate the danger. A once isolationist America learned that inaction leads to intimidation from other countries, as happened with Japan, Italy, and Germany in rapid fire sequence during the 1930s. From Pearl Harbor, America learned that its oceans no longer protected it from aggressive despotisms abroad. Thus, America dropped its previous strategy of isolationism as unrealistic in light of advancing technology. On September 11, 2001, America discovered the danger of well-financed international terrorists not directly controlled by one nation. They can strike anytime and almost any place, using our open democracy against us. At present only America has the global reach and power to mount a counterattack. The United Nations was as helpless when confronted by the crises of the Balkans, African tribal warfare, and Iraq as was the League of Nations was when confronted with the aggression of Japan, Italy, and Germany.

Conflicting interests among the more numerous UN members necessarily induces extreme caution, but a caution that emboldens bullies. Iraq suggested the Western Alliance (NATO) is nearly dead. Moreover, the United States, not Europe, is the main target of international terrorists, along with Israel. Many Americans now suspect that some Europeans may even secretly encourage and support them. America wants world peace and justice, but not necessarily at any price. Many Americans argue that history shows that peace comes through strength, not through the appeasement of zealots who have little sympathy with its worldview.

Tribal and religious wars seemed to have largely been eliminated or at least suppressed by European imperialism. Yet, once the imperial powers left a region, tribe-like warfare promptly broke out, and it has become worse since the end of the Cold War. Without a central authority, that tribal warfare emerges almost spontaneously. When populations are small, such warfare tends to be relatively low key, sometimes merely symbolic. But the larger the population, the more murderous tribal conflict can and usually does become. Saddam Hussein's massacre of the Kurds with poison gas is a good example. Early despotisms emerged as a means of quashing violent anarchy within tribes. The early empires that soon followed aimed to quash anarchy between tribes.

Tribalism and the almost constant strife it brings with it is thus a long-standing problem, a lasting legacy of the Agricultural Revolution. Farming created a new ecosystem that provided humans with ecological release from the strict limits and constraints on population growth imposed by a nomadic hunter–gatherer ecosystem. But the crash of murderous anarchy soon followed, as we have repeatedly stressed. In the political adaptation stage that came next at the dawn of civilization, local despotisms and then empires emerged to quell the anarchy of tribalism. But tribalism struck again after an exhausted Rome could no longer restrain tribes bordering its territories. The Mongolian tribes came out of central Asia to overrun both China and much of Islam, pushing into eastern Europe and putting many to the sword as they advanced. With the rise of (for the times) a high-tech European imperialism with a global reach, tribal warfare seemed once again on a sharp decline. Today, the imperial powers are gone, and the Cold War is over. So once again the world is faced with resurgent tribalism made all the more deadly because of a three-fold increase in the world population since the end of World War II. Additionally, the warring tribes now have access to a massive array of far more murderous conventional weapons than ever before as well as the nuclear, chemical, and biological ones. Tribal leaders rarely have hesitated to use those weapons when they can. Thus, tribalism itself has become an incentive to continue to invent, produce, and distribute these weapons. The more that rogue states or terrorists acquire high-tech weapons, the greater the need to invent even better ones, and so on.

This is "the new world order," a high-tech world of fragmented nations, international terrorists who see themselves as "freedom fighters," lawless mobsters, and resurgent tribal strife awash in anarchy. The prospect of this world coming together in a spirit of goodwill and cooperation to solve mutual problems seems a bit remote, absent a common and over-arching threat.

Tribalism has been brought under control historically through strong and often ruthless application of force by a transcending imperial power. Still, ruthless imperial dominance is a strategy that flies in the face of everything liberal democracies stand for including such high-tech cowboy democracies as America. Tribal terrorists know that and they take full advantage of it.

22

TECHNOLOGY AND THE GLOBAL ECOSYSTEM

Technology itself does not threaten an ecosystem but its application can and often has on a local level. Even the first humans's primitive existence already had a dramatic effect on the ecosystems wherever they roamed. Primitive people wiped out much of the mega fauna in North and South America, Australia, and all the islands of the Pacific thus to change those ecosystems.[1]

Only because of advances in technology can more than six billion people live on planet earth today, mainly in cities, towns, and suburban sprawls. First World people lead sedentary lives for the most part, spared much physical exertion because machines burning mostly fossil fuels do nearly all the work. Industrialized nations require a global network of railroad lines, airlines, and highways, augmented by telephone lines, cell phone transmitters, and orbiting satellites. Airports girdle the globe to accommodate the airlines flying millions of people every day. Five hundred years ago, crossing the Atlantic Ocean took six weeks and now can be done in six hours. High-rise buildings now sprout up from nearly every large city in the world. Adding what each person now consumes and taking into account all this infrastructure—the schools, hospitals, central power stations, factories, production equipment, and trucks—each person now accounts for about a hundred times as much consumption as his or her late Neolithic counterpart of ten thousand years ago.

The global environment has been drastically modified in the process of human evolution and its more sophisticated techno-structure. Yet this is not new: Life has been in a constant process of changing the global environment since life first surfaced four billion years ago. Humans have influenced the earth's biocosm from about forty thousand feet down to forty thousand feet up. From the third dead rock from the sun, the earth has evolved into a planet teaming with life and, in the view of some, the whole of life evolved into a living entity called Gaia.[2]

The explosion in human population and its higher per capita consumption, however, began just ten thousand years ago. If we compress human history into one year, this explosion began on the afternoon of day 365, and about two thirds of that happened since 11:30 p.m. on that day. Such is the applied power of today's human techno-structure.

In the worst view of this progression, humans have morphed into a rapidly metastasizing cancer and have sucked the substance out of their host environment, thus threatening to kill it. The seventeen Ice Ages probably created the disruptions necessary for FROCA to trigger the bioevolution of large brains that enabled people to create awesome technology. Pessimists might argue that our braininess was nature's primary evolutionary mistake.

Such a dismal view surfaced in the last fifty years when human consumption took a sudden quantum jump. America's continuous wave of new technological frontiers had traditionally been a great source of pride. Now that passion for new technology is often seen as a toxic temptation that has a life-threatening downside. Still, not many people take the extreme view seriously that human technology will destroy the earth. No catastrophe yet has.

However, continual accelerating, and largely unconstrained, application of ever newer technology could culminate in a final ecosystem crash. That crash might reduce human population by upward of 50 percent. This is not just an alarmist view. In the course of our evolution, many local kill-offs and extinctions have taken place. The Black (Bubonic) Plague killed more than 30 percent of Europe's population in the fourteenth and fifteenth centuries. Smallpox killed upward of 90 percent of Central America's Amerindians in the sixteenth century, and similar kill-offs took place in Australia and North and South America (such as the cholera epidemic). Little Easter Island, as a microcosm example, lost as much as 90 percent of its population directly as a result of its own technology and internal divisiveness, not an invasion by a foreign virus. All the hominid species became extinct except for *Homo neanderthalensis,* modern humans's near cousins, the Neanderthals. We don't know exactly why other hominid lines became extinct, but they did, and that is the point.

Meanwhile, the application of human technology has caused hundreds of well-documented catastrophes. In the Americas, the ancient Clovis peoples eliminated the North American mega fauna through over-hunting. Modern American Midwest farmers draw down the Ogallala aquifer at an alarming rate of 1.74 feet per year to achieve its huge output of monoculture grains; this one aquifer irrigates at least one fifth of all U.S. cropland. Mexico, like other nations worldwide, faces its own water problems. Today's Mexicans must learn to both conserve water and use it responsibly, which will take education and awareness that water is an expensive resource. Culturally, Mexico's growing population has not accepted the real costs of water consumption and sewage services. A water crisis may strike the country within two decades. The world's tropical rainforests are being cut down at a rate of 149 acres per minute, and they have a limited capacity to recover because of heavy rainfall and soil leaching. Air pollution now kills three times more people worldwide than do traffic accidents.

Further, global warming marches on and comes with two different and indeed opposite disaster scenarios. In scenario one, as the globe warms up, the glaciers and ice caps melt, the ocean levels rise as much as 20 feet (6 meters), flooding coastal plains around the world that account for much of the world's food supply. The Netherlands, Denmark, Singapore, much of Egypt, the southern U.S. states such as Florida and Louisiana, and islands everywhere would be underwater or nearly so. London, New York City, Tokyo, Shanghai, Los Angeles, and indeed most coastal cities would need to be evacuated, or at least partially so. Some rivers are overflowing because of melting glaciers while others are drying up because of accelerated evaporation. Disease and drought decimate some crops while the raised levels of heat cause others to grow faster. However, a rise in sea levels around the world would flood enough farmland to kill billions of people through starvation.

The second disaster scenario brings us to the new Ice Age, paradoxically triggered by global warming. To briefly review, the Arctic Ice Cap and Greenland's ice cover have been melting rapidly in recent years. The North Pole now often sits over open water and not thirty feet of ice as it did sixty years ago. Melting ice caps release fresh water that dilutes the ocean's salt content in that region. Greenland, a huge island, contains the world's largest fiords as well as an ice sheet once up to 10,000 feet thick. As the ice melts, chunks of glaciers can break off and block the entry or exit to the fiord. As the ice converts into fresh water, lakes can build up behind the ice dams. Some of these fiords are situated near where the Gulf Stream moves south, warming the water again before returning north. Water movement happens at the bottom of the ocean (not on the surface) because the water is heavy with salt. The prevailing winds blow across the Gulf Stream all the way to Norway and as they blow, they draw off water and heat but leave the salt and colder water behind. Both the heat and moisture fall on western Europe, giving it a much milder and wetter climate compared to its latitudinal counterparts of Canada and Siberia. Europe's wet and mild climate allow it to grow enough food to feed about six hundred million people. Canada, by comparison, is about the same size as Europe, but can grow enough food to feed only about thirty million people. If Greenland's ice dams, holding back huge fresh water lakes in its fiords suddenly broke, they would release a wall of water into the ocean at the spot where salt-heavy downwelling takes place. The result, according to some analysts, is that the down-welling would shift south by several hundred miles, thus foreshortening the Gulf Stream. No longer would moist wet winds blow across Europe. The European climate immediately would become much colder, the growing seasons much shorter, and farm output greatly reduced. Europe could then feed less than 20 percent of its population.[3]

The earth has experienced seventeen Ice Ages lasting on the average of about a hundred thousand years in the last two million years. The warm interglacial periods last about twelve to fifteen thousand years. The interglacial we now live in is already about fifteen thousand years old. These Ice Ages did far more damage to preexisting ecosystems than anything we humans have done. True, they didn't destroy our cities or farms because these things had yet to evolve. Yet if the above scenario does occur, humans will almost certainly have been at least part of the cause.

While global warming is clearly related—at least in good part—to the broad spectrum of human technology, other environmental disasters are also possible. A recurrent concern is of a cascading chain reaction arising from what might have been perceived initially as an insignificant accident. A fictional story that can illustrate a potential extinction event is a major plot line in the 1992 Clive Cussler book *Sahara.* Scientists discover a deadly, mutant form of red tide (toxic algae bloom) arising off the mouth of Africa's Niger River. It is propagating far faster than normal and it is picking up speed. Besides being toxic, it consumes the oxygen in the water, thus suffocating fish and other marine organisms. Since about 70 percent of all new oxygen comes from diatoms that live in the sea and the toxic tide is killing them, this continuing process would also kill off humanity in about two years. Cussler's hero, Dirk Pitt, is given the job of going up the Niger to track down the cause. He does, after the obligatory hair-raising adventures. Just in time (of course) the world is saved from disaster after the necessary correctives take place. The problem was caused by a French controlled nuclear dump site in Mali that bled radioactive waste via an aquifer into the Niger.

We can't possibly say for certain some similar disaster won't strike, nor can we expect to anticipate every possiblity in advance, no matter how careful we are. Chaos theory holds that non-linear relationships often cause a collapse of order into chaos without any warning. With respect to technology, we are also in a long-standing bind. As James Burke and Robert Ornstein point out in their 1992 book, *The AxeMakers Gift,* ever since the first hominid crafted the first hand ax, humans have become ever more dependent on technology. Our so-called cultural progress has coevolved with our technology. Burke and Ornstein suggest that each new bit of technology seems to "make us an offer we can't refuse." It solves a problem or meets a need and so we adopt it, become dependent upon it, and consequently learn about its accompanying disadvantages after the fact.[4]

Technology createth and technology taketh away. Technology has brought on a host of environmental problems, yet later technology often clears them up. For example, the early technology of iron making required wood in the form of charcoal that came from cutting down the forests of England. Before long iron makers (among others) had nearly exhausted those forests. In the seventeenth and eighteenth centuries, coal was first chipped out of shoreline cliffs for ships. Underground mines were first operated by the British Empire to further furnish its ships' need for coal. England's forests then came back. When the steam engine was invented to pump water out of the coal mines, coal replaced wood for home heating. Home coal fires became the source of London's smog of the period 1850 to 1956. That smog blackened all stone buildings and caused a sometimes fatal form of air pollution. An estimated, thirteen thousand Londoners died from that smog during three weeks in 1952. Therefore, coal was banished and was replaced by cleaner burning coke. London air is now far better, its buildings cleaned up, and "fogs" are comparatively rare. The automobile traces a similar history as earlier noted.

On a broader perspective, however, "hard greens" (people who want to protect the wilderness, oceans, rivers, lakes, and shoreline) point out that the population explosion surely was an outgrowth of technology.[5] Modern technology, however, can correct and reduce the birth rate. Birth control pills and other contraceptives are only a minor part of this. Larger reduction comes from broadly based affluence. First World affluent people have fewer children than Third World people do. As a result, future population growth is predicted to rise fastest in technologically less advanced countries. Some estimates project that the world's population will reach nine billion by 2050 and then start to decline.[6] This is based on the fact that birth rates have dropped to the point that the most affluent nations are hardly reproducing themselves. The populations of many European countries are already beginning to decline, even Russia's if for somewhat different reasons. Japan barely reproduces itself. The population of the United States is growing mainly from immigration. Only in the world's poor nations does population still rise at rapid rates.

Moreover, the hard greens point out, only the affluent nations take pains to protect the environment, pass strict protective legislation, and plan for the future. Poor nations are too busy trying to feed themselves to worry about the environment.[7] For instance, Brazilians sometimes respond to complaints about the disappearance of the Amazon rain forests by saying, If you think it is that important to protect the rain forests, then pay us to do it.

To feed the world's population, monoculture, chemically treated grain production may not result in superior nutrition. However, simplified technology can be used to serve humans in the changing environmental conditions that we helped to create. The Rodale Institute's Farming Systems Trial, which began in 1981, is one of the longest running crop sustainability comparison experiments in the world and demonstrates not only how the organic system gives better yields of corn and soybeans under severe drought conditions but also how the organic system allows for higher environmental stability under flood conditions because of less runoff and harvesting more water for groundwater recharge. The concept of "regenerative agriculture" has developed into its own science field over the last twenty years, involving low-tech farming, as well as focusing on regenerating woodlands and streams. Thinking in terms of regenerative agriculture harkens back to the importance of farmers in society. Agriculture is the foundation of all civilization, and improvement in agriculture elevates the whole society through a cleaner environment and better health.

Meanwhile, the Gulf Stream problem alluded to earlier could be prevented by the application of the appropriate, existing technology—provided plans to do so were made in time. Recent movies feature space-age technology as a way to thwart a future asteroid or comet hit on the Earth, again given the proper planning and lead times. Fortunately, many of the computer models that once predicted disaster were deeply flawed by an inherent characteristic of all models. They can forecast "objectively" only by assuming past trends continue in a linear fashion. If nonlinearity is assumed, then a multitude of outcomes can emerge from the same algorithms.[8] Outcomes then depend on guesses. Many such linear models failed us in the past because so much of the natural world operates along nonlinear paths. Chaos theory takes account of nature's nonlinearity and says that this fact makes the future indeterminate, at least in detail. Thus, we cannot confidently and accurately predict the timing of a disaster, if it comes. It is much more likely to take us by surprise. (In December 1941, the American government knew that war was about to break out. Still, Pearl Harbor came as a terrible shock.)

Most predictions of future technology suffer the same fate. Despite, for example, all the predictions for new technology coming out of the New York and San Francisco World Fairs of 1939, no one predicted atomic power would emerge six years later or that Bell Labs would invent the transistor in the next decade. No one anticipated electronic computers or commercial jets, although they foresaw television's future popularity because it was already in limited use.

In another example, the predictions of the 1970s called for bone-dry oil wells before the year 2000. Instead, thanks to improving technology in oil drilling and conservation, and later discoveries of oil wells that forecasters had no evidence for at all, the world's proven oil reserves remain high. On the other hand, China's rapid economic growth has put huge new pressure on oil supplies.

Better technology may save us from our self-created problems and natural disasters or it may not. The preponderance of evidence does not decisively favor either theory. Still, it might not much matter if the preponderance of evidence swung decisively in favor of the meta-crash for technology as a whole. Given today's fragmented and retribalized world, it would be very difficult to do much about it. Democratic governments resist proactively taking controversial or risky action if reasonable doubt exists about the danger. People can always find reasonable doubt if they look hard and creatively for it, right up to a crash. Every speculative bubble comes with skeptics who predict a crash is coming, but they are usually ignored by those who argue that new frontiers have created new rules that will preclude any future failures. Additionally, most of the dire predictions are for the relative near term. Often they prove wrong thus encouraging speculators to keep a boom going, again, because "this time is different." The recent crash of 2000 to 2002 was preceded by exponential growth in the market that continued to rise and led people into believing that "this boom is different."

In short, we will know we have had a meta-crash only when it arrives. To proactively or preemptively shut down new technology (and take a huge hit in unemployment or in real income) before the crash would be all but impossible in today's world. There is no authority that could order—let alone enforce—a shutdown, certainly not the United Nations. The effort to create such an authority would meet with dogged resistance by all nation states . . . before the crash. When President George W. Bush moved "preemptively" into Iraq, the outcry was deafening. Democracies cannot arrest known murderers or rapists merely on suspicion they may rape or kill later. They can be arrested only after they do rape or kill again. The same restraint applies to technology that might rape the environment or kill many of us. Nations can take politically acceptable steps to constrain technology effectively (that is, comprehensively) only after its excesses strike many of us dead, and not always even then, if history is any guide.

23

A POST-CRASH THEOCRACY FOR A STABLE GLOBAL ECOSYSTEM

This chapter takes a purely speculative look at possible future "software" humans might create that brings about stable cultural and ecological equilibrium following a meta-crash. I do not predict such a crash nor do I think one is inevitable. Still, my scenario for a stable eco-future does draw on historical experience about how past cultures brought technology under control. History suggests, as we pointed out in the last chapter, that before humanity is likely to take the preemptive action to forestall a future crash, we must first feel the need to take drastic action beyond all reasonable doubt. To experience a crash first is to remove all reasonable doubt.

Recall that despotic governments evolved to quash spontaneous local anarchy. This anarchy emerged from population growth after the constraints on such growth were released by new agricultural technology about seven thousand years ago. The steps toward despotism came from the bottom up but soon became top down as explained earlier. It began with elected headmen who later became chiefs, then hereditary chieftains, then kings, then a god-king. Kings came from a single (royal) family line or dynasty via inheritance. This method might not produce consistently good kings, but a mediocre prince was preferable to fighting a civil war that would decide who would be the next chief. That policy aimed to preserve order.

What had been bottom up now became top down. God-kings were absolute monarchs and despots. The evolution from egalitarianism to a socially stratified and slave-based despotism did quell anarchy by imposing law and order. However, anarchy had only been pushed to the borders of the state. At these borders, conflict continued between tribes rather than within them. The usual patterns of tribal conflict included raids (often to obtain slaves) followed by counter raids, as well as revenge killings, long-lasting feuds, and battles to obtain scarce resources.

This borderland anarchy could seriously threaten trade. Trade was increasingly vital to the emerging states because their much greater division of labor was a hallmark of civilization. Tribal conflicts could also deny a state access to vital resources. Civilizations require public buildings and thus building stone. The river valleys where the cities arose often had no stone quarries. When the king sent out parties to mine the stone and bring it back, the transportation routes were often blocked by rivals who demanded tribute for safe passage. Costly wars often broke out.

Strong despots countered by enlarging their domains. They created empires that embraced a variety of tribes, chieftains, and even smaller states. Again the idea was the same, quell anarchy from the top down, and by brute force if necessary, to maintain a stable social ecosystem in stasis. The impulse to subdue competition was and is very strong. If competition becomes uncontrolled, anarchy threatens and can spell ruin for those in power. In 1971, Lionel Tiger and Robin Fox wrote *The Imperial Animal*.[1] Humanity's "imperial impulse" is an impulse toward stasis, something all species strive for in a variety of ways. Only humans do it self-consciously, whereas other species do not. The imperial impulse aims to create a stable social ecosystem, by compulsion if necessary. Political empires, business empires, religious empires all aim to expand central authority to encompass and then quash any competition that threatens to disrupt the stability of its power. The early empires of Egypt, Babylon, and later of Persia, Greece, Rome, China, Aztecs, and Incas all had this goal in mind.

All ancient empires were caught in conflicting and competing requirements. The system that enabled them to gain enough control to maintain stability also put a damper on technological innovation because new technology could disrupt control. Thus, early empires were limited to the technology they had at inception. They possessed enough technology to create and sustain civilizations built around states, and to maintain order within those states. They also were able to expand those states beyond their borders and thus create empires. But that technology was never advanced enough to expand an empire very far and maintain control. Horses organized as cavalry made it possible to push out quite far (witness early Islam or the Mongols) under the right conditions. Yet having successfully acquired a geographically large empire, political leaders found it nearly impossible for a central authority to maintain control. Transportation and communication were inadequate and slow. The empires of Alexander, Rome, Islam, and Genghis Khan all broke apart as a result. The faster an empire expanded, as a rule, the faster it tended to fall apart.

Still, the imperial impulse was to create an empire encompassing the "known" world and put it into one social order. Technology, however, was much too weak in part because it was hobbled by the chains of despotism. No empire has yet encompassed its entire known world although Rome came close, and in a way, Britain came even closer by circling the globe, but Britain was far from controlling all of it.

Today, technology is in place, particularly in transport and communications, to create an effective central world government. We also have the military means to impose control by brute force if need be. Ironically, our democratic political system based on individual rights and free markets would prevent that. Democracy allowed for and protected the innovators who created our most advanced modern technology and rejects the idea of a central world authority with enough power to quash all serious resistance. Moreover, the democratic system extols the advantages of competition in business, government (via elections), and team sports. It protects competition within the free market system by constraining the natural tendency to create trusts and cartels in the interests of stasis. Left to itself, market competition ends with monopoly capitalism operating on the same logic and imperial impulse that led despots to create empires. Thus anti-trust laws ensure market competition. Organized athletics gives the weaker teams the first choice of draft picks to prevent any one team from becoming all powerful. Poor golfers are handicapped so they can compete with the better golfers.

Today, the technological flesh is plenty strong enough to impose central control on the world, but ironically democracy has weakened the imperial spirit too much for it to try. The twentieth century saw the end of two formidable efforts to create world empires (of fascism and communism respectively). Both ideologies took a jaundiced view of competition. Communism ruled it out as a matter of principle. The very system that corralled the imperial spirit that aimed for World Empire could hardly be expected to revive it. The whole idea of democracy implies global pluralism among the nation states and thus permits many variations on the democratic theme. Democracy also implies, however, that the nation-state is the focus of the citizen's loyalty rather than one's ethnic, tribal, or religious identity. Democracy also encourages competitive spirit rather than attempting to suppress it. The free enterprise system allows us to express those ideas in the competitive marketplace. Competition is expressed in politics with elections that force candidates to compete for office and for political parties to compete for both administrative and legislative control.

However, much of the world's population is not comfortable with so much competition on so many levels. Democratic competition implies a frontier type of life, a life of dynamic instability that many people don't want. Encouraging political democracy in these regions often backfires and can actually promote tribal conflict. Democracy does not compete well with established local political systems. Similar impulses, if unchecked, soon propel market capitalism into monopoly capitalism. Unless constrained from doing so, the stronger drive out the weaker, or in the case of an impasse, the strongest enter into cartels that control prices and divide up territories among the strongest oligarchs, who then set strict limits on the competition between them and work together to freeze out new entrants.

What if our ongoing frontiers of new technology also conform to the FROCA process? What if, as a whole, technology overexploits the opportunities it creates and then crashes? What if the explosion of consumption it fosters causes a sudden serious collapse of a global ecosystem like the one we described with the shift in the Gulf Stream? That scenario qualifies as a global meta-crash that creates a new anarchy by inundating local authorities with cold, starvation, and desperate refugees from a newly frigid Europe. A democratic solution would be unworkable in such a crisis. A demand to restore order by any means would likely erupt from the bottom up, a demand to restore order by force if necessary. People have often voted despots into office hoping they would become a stabilizing influence. Still despotism is a far cry from the easy egalitarianism of the hunter–gatherer bands.

So, why did people historically turn away from an egalitarian lifestyle and accept social stratification, slavery, and huge disparities in wealth and income? They may well have had an imperial impulse to seek safety from the murderous anarchy of large ungoverned populations. Democracy, however, taps into an equally potent impulse toward egalitarianism. That impulse rejects the kleptocracy that seems to seduce despotisms and in the end becomes a toxic temptation of those in power, finally destroying them. Out of those ashes democracy arose.

One answer is that religion helped to make despotism acceptable in nearly all ancient civilizations, including the kleptocracy that went with it. (In the communist and fascist nation-states, ideology played a similar role of justification.) Most despotisms arose from tribal religion. We know of no early state that had a secular government. The chiefs were often shamans as well. Religion helped to unify the state and its people. Religion made the sacrifices the people were called upon to make to serve a higher divine purpose. Despotism provided greater safety from anarchy although it came with a high price.

The great bulk of the people had to accept a low social status and poverty amid the wealth and luxury of the despots on high. Full-blown and kleptocratic despots arrived not as chiefs, but as god-kings. Most people could easily believe that god-like people *should* live to a higher standard than mere mortal humans. Kleptocracy thus lifted the god-king above the masses of ordinary people, who would then stand in awe of him. To submit to the head of state was also to submit to the gods.

This brief overview suggests three preconditions to a new order that can constrain new technology in the interests of a safe and stable global ecosystem. First, we would actually need to have a meta-crash severe enough to remove all doubt that something drastic must be done NOW. Second, a critical mass of survivors of the crash would need to accept submission to a single central authority that has the power to act decisively—and ruthlessly if necessary—to impose its will. (Looters would be shot on sight; trial lawyers would be rounded up to form compulsory eco-labor battalions.) Total chaos and a wide-spread breakdown of local law and order might convince people that an authoritarian government is the only practical choice. The longer the preceding period of chaos, the less doubt of the need.

Third, once order is restored, the general population would need a reason beyond safety to accept a post-democratic central world authority as the new standard. Removing the immediate danger would not be enough. For example, at the outset of the Civil War, the people in Washington, D.C., accepted President Lincoln's edict of draconian martial law and the suspension of habeas corpus for most of the war. The Confederates were too close and well infiltrated into the population. But as soon as the war ended, people promptly demanded, and received, a full return to peacetime civil rights. In the hypothetical case, an obvious and compelling reason would need to exist to maintain what amounts to martial law long after immediate danger has passed.

For many, protecting the environment—Mother Earth, Gaia—from further damage could provide that reason. Sacrifices in the interest of the environment would need to be made and accepted as legitimate on a continuing basis. In short, people would need to find some higher spiritual justification to accept those sacrifices and constraints on freedom and liberty. For example, people may understand that individual liberty turned into selfish license that created the environmental problems. Many people already claim that our own material excesses pose a great danger to the global ecosystem. If a great meta-crash did occur, that fear would cease seeming alarmist and become mainstream. The danger of excessive consumption would become a key tenet of the eco-theology that justifies an eco-theocracy. Personal freedom, privacy, and civil rights, including the right to innovate, could cease being democratic virtues. They could well be seen instead as vices that produced ecological disaster followed by anarchy and chaos. A Bill of Personal Eco-obligations to the Environment might replace a Bill of Personal Rights.

No existing religion, by itself, seems able to support this change. All have too much tribal-specific or ethnic-specific history. Still, current religions contain some elements that a future ecological theology could use. An eco-faith would focus on the preservation of the globe's complex and fragile ecosystem, and humans and their offspring would have a direct and obvious stake in that. The meta-crash itself would have made the danger crystal clear. Pointing toward a possible future Armageddon does not. History is indeed fraught with doomsday predictions that people routinely ignore because they infrequently come to pass. Armageddon must actually strike to convince us of that we suffer the consequences of our own past sins. The stockmarket crash of 1929, Pearl Harbor, the dot-com bust, and September 11 terrorism all were crashes that led to changed behavior but only after and not before those events. In short, when disaster strikes we often turn away from past "sins." In a meta-crash we might seek redemption in a new and entirely different and eco-friendly way of life as spelled out in a new eco-theology.

Accordingly, the eco-faith needs a detailed set of rules to live by. The rules would in effect force everyone to remain mindful of treating the environment and its ecosystem with deep and sacred respect. This eco-faith thus could draw from both Judaism and Islam for the idea of daily eco-friendly rules and rituals. It could draw from the Amerindians and other animists' traditions that focus on a deep respect for nature. Chief Seattle's famous message could become a new prayer.

Meanwhile, the post-crash ecosystem would need an immense amount of remediation work paid at a minimum rate. Thus, the work ethic embedded in both the Benedictine and Calvinist traditions, with suitable adjustments, could play a useful role. Indeed, service in labor battalions working on remediation might well become a required rite of passage for young people. They would learn ecology by doing the work, not by studying the books. Some people might choose to become eco-monks who make ecological service the mission of their lives. Monks could glean spiritual sustenance from such eco-work that would be its own reward, and frugality in consumption would be an integral part of their spiritual journey.

Sumptuous living would become an eco-sin and an eco-theocracy would discourage material wealth beyond a fairly modest level. We must learn to live in frugal harmony with nature and each other. The human footprint must be kept as light as possible. Whether or not that idea would be enough to prevent kleptocracy is open to question; early Christians after all were supposed to live a similar way and they often did at first. But when the Church became the official church of the Roman Empire with much discretionary power, the Christian Church soon became a religious kleptocracy where top church officials lived in luxury amid abject poverty.

A passage from Genesis might have a place in an eco-theocracy: "Of every tree of the garden thou mayest freely eat: But from the tree of the knowledge of good and evil, thou shalt not eat of it: for in the day thou eatest thereof, thou shalt surely die." Adam and Eve violated the tree of knowledge when they bit into its forbidden fruit. That bite launched humanity down the road of knowledge—ascending technology—and that ends in a meta-crash. Look where this passion for learning and knowledge got us, the eco-priests might say. Pursuing knowledge simply has led humans to eco-perdition once we became hooked on the toxic temptations of our own technology that we express as a consumption binge.

The hard green philosophy of withholding huge tracts of Mother Earth from human habitation for the sake of ecological preservation could be invoked. Economic growth would no longer be a goal, nor would ever-rising material standards of living. Again, those were the mistakes that cause meta-crashes. It would, however, be necessary to create a place for everyone, then make sure everyone kept that place, by force if necessary. The eco-way of life must discard most of the modern world's affluent extras onto an eco-junk pile as we celebrate a new spirit of frugality. We could all eat organic food exclusively if the meta-crash reduced much of the population and the government imposed strict limits on population growth.

Perhaps I've sketched out enough to show such a drastic shift is not likely to happen without the overarching incentive provided by direct experience, not just a prediction. In short, it would take a very big crash indeed to convince people of the necessity for such a drastic shift to an eco-friendly way of life. My guess, and it is purely a guess, is that it might take a kill-off of about 50 percent, perhaps more, to set the stage. Even then it might take a period of murderous anarchy to drive the point home. To repeat the Spanish proverb, "Better a thousand years of tyranny than one day of anarchy."

The broader psychology behind this chapter is familiar to people who have tried to battle addictive behavior. The individual must have no options left before he or she can break out of the denial of addiction and make the drastic lifestyle changes necessary before recovery can take place. I am suggesting the same process applies to communities of addicts. Perhaps, as soon as we humans became dependent on technology, we were lost; it could only end in ruin. In this chapter, I assumed (not tried to prove) that this ruin does indeed take place in some unspecified future and humanity actually allows its technological options to run out. In short, I imagined a post-crash scenario in light of our past cultural evolution via the FROCA process. I have conjectured a broad pattern of post-crash adaptation that is the final "A" stage of FROCA. It is important to remember that this scenario from the dark side, while it might well happen, is by no means inevitable. Life always evolves on the edge of chaos; crashes have been part of that chaos. We can indeed thank past crises, the growth followed by crashes that came with them, for what we are today.

24

A MORE HOPEFUL SCENARIO

Humans have evolved as the only high-tech species on earth. We are nature's natural technologists, and that same nature gave us the power to take control of our cultural evolution. We may soon take control of our anatomical evolution as well, via bioengineering. Human inventions in technology are as much a part of the natural world as is a bird's nest, an anthill, or a beaver dam. Life has evolved to ever more complex levels and humans can now create ever more complex techno-structures that allow us to control much of our own habitat and its ecosystem. It does not follow that our evolutionary dynamic *must* inevitably end in a meta-crash caused by humans overexploiting our ever newer technology.

True, every positive self-reinforcing feedback loop must either stop on its own or else crash. But who is to say that humans cannot slow down the loop before it brings down some vital part of the global ecosystem? We have a long history of technological applications that caused damage to the ecosystem. Still, later technology corrected some of the problems and that very process has been a part of human cultural evolution. We have no compelling reason to assume that the process of self-correction cannot continue. Some environmentalists respond to this point by saying, "Okay, let's continue to let the environment run that experiment." Many technology-induced general disasters have been forecast, but as yet none has come to pass.

Meanwhile, constraining new technology courts it own risks. A ban on technology would not protect us from future asteroid hits or massive climate shifts. Such past disruptions have come often independent of human technology. Yet our technology could possibly avert the worst part of such threats. We have already pointed out that in less than two million years, the Earth has had seventeen separate Ice Ages, some with mini-cycles within them. Based on statistics, we are now due for another Ice Age. It was our technology that allowed us to survive the Ice Ages and even embrace the colder climes. When the next one arrives, we may need all the new technology we have developed to cope with it. During past Ice Ages past, new technology gave humans the ability to make fire, warm clothing, and construct shelters of various kinds. In the process of becoming pioneers on these new frontiers, we often discover we can exploit new opportunities. True, we have a long history of overexploiting local frontier opportunities. Such overexploitation will continue, followed by some level of crash. Yet new technology will not force the whole Earth to crash.

Human culture has experienced countless crashes big and small, some with major kill-offs. Those crashes make up a crucial part of life's trial and error pattern of evolution that allows us to learn from past mistakes. We learn from the crash and survive by adapting to a new cultural ecosystem that usually corrects for the excesses of the immediate past. Moreover, it would seem reasonable that, if we better understood the FROCA process, we would improve at curbing some of the worst excesses.

Biotechnology and genetic engineering are beginning to make an impact. Francis Fukuyama writes in *Our Posthuman Future: Consequences of the Biotechnology Revolution* that we might so redesign our own DNA that a posthuman creature will emerge.[1] If we do learn how to redesign our own DNA, it would be the ultimate triumph of human cultural coevolution. Our technology has already preempted random mutations in human DNA as the major force for change at least for thirty thousand years. Now we seem poised on the brink of taking over biological evolution by replacing random mutations in our DNA with genetically engineered changes. Before long our progeny would not be human; they would be posthuman, as Fukuyama points out. However, all humans are post-hominid, and we could call the hominids post-apes. In fact, all multicelled creatures are post-bacteria. Evolution since the hominids, however, is a story dominated by our own innovations in technology. Had the hominids not invented their Oldowon stone tools, it is doubtful we would have acquired a brain capable of discovering DNA let alone taking control of it. As Isaac Asimov wrote in *Life and Energy,* cultural evolutionary breakthroughs include four milestones: use of tools, control of fire, application of the steam engine, and control of nuclear power.[2]

Moreover, as soon as we began the shift to agriculture, we began reshaping the DNA of both domesticated plants and animals. We reshaped the DNA by trial and error. When a mutation that we liked appeared, we reproduced it. We exerted our control via selective breeding and cross breeding until we acquired the characteristics we wanted. While random mutations still take place, humans decide which random mutations will be judged fit enough to survive. Most of what we eat and wear is genetically reengineered via cross breeding plants and animals away from their native versions, and this has been going on for some millennia.

Should we take that next step? Should we permit bioengineering to replace the random mutations in our own DNA? If the answer is no, no global political power exists to enforce that decision. In the last chapter we briefly looked at what it might take to put such a political power in place. Individual nations might ban bioengineering, but if they do, they make themselves vulnerable to rival states that continue with bioengineering in hope of achieving competitive advantage. In 1945, the United States developed the atomic bomb out of fear that the Germans would develop it first and had hope that it would remain a U.S. monopoly. However, the Soviets obtained nuclear weapons in the 1950s, followed by Britain (1952), France (1964), China (1964 for atomic bomb, 1967 for hydrogen bomb), India

(1988), Pakistan (1998), North Korea (2004), and, possibly, Israel and Iran. If a new technology comes along that seems to convey an important competitive (survival) advantage, it will propagate because no central world authority exists now and never has. That is why technology keeps on propagating through competition.

Today, we live in a post-imperial, post-Cold War, and retribalizing world that seems increasingly anarchic. It is a dangerous world now that such lethal weapons are at hand and so widely distributed. Unless a new general threat to the West arises, the Western Alliance appears, if not exactly dead, too weak to bring order out of this chaos. It works, when it works at all, mainly from a consensus that will rarely appear in the absence of a clear and present danger to everyone. The United Nation suffers the same problem to a much higher degree because the membership is larger and has far more diverse and seriously conflicting interests arising from conflicting worldviews.

When interests do conflict, we discover what seems to be enormous intransigence in the course of achieving durable diplomatic solutions. What we are witnessing is an aspect of life that impels everything from bacteria to cells to tissues and organs to organisms to communities (or cultures) of organisms to act self-consistently in an ever-changing environment. Autopoiesis plays a major role in the ability of all life to self-organize and at all levels. Autopoiesis drives our bodies to heal themselves if injured or sick. It drives communities to restore order after a flood, earthquake, or similar disasters, up to certain limits. Yet it is a force that impels job defensiveness in bureaucracies worldwide while also encouraging the formation of team spirit in membership organizations such as true partnerships.[3] Autopoiesis, in other words, can act in both positive and negative ways.

Autopoiesis mimics the "law" of self-preservation, a law most people accept. Paradoxically, it can also account for individual or even mass suicide. The precise behavior autopoiesis induces in humans depends on the nature of the identity that the individual or community has of itself. Where people identify closely or deeply with their traditions and religious or ideological values, they will often defend that identity to the death. That was indeed what Japan's kamikaze pilots did toward the end of World War II. To them the survival of Japan and its Japanese spirit, with which they closely identified, was an identity they could try to save only through self-sacrifice. We have seen similar behavior for similar reasons by Islamic suicide bombers.

In short, autopoietic intransigence arises from threats to our core identities, a threat to our very soul. Such conflicts are rarely resolved by negotiation, except under intense outside pressure to reach a compromise. Otherwise the contestants might withdraw from each other or agree to an armed truce. But each side retains full possession of their core identities that give meaning to their lives and for which they are often willing to die. Intransigence allows cultural groups sharing a core identity to survive in the face of the cross currents of cultural conflict that are part of life today in our newly created electronic global village.

Geographic separation can no longer keep the peace. But intransigence is a way of creating psychological separation. One becomes so adamant about one's beliefs that outsiders see that there is no point in arguing about it. One achieves a psychological separation in the absence of physical separation. The rise of fundamentalism in most religions reflects this. Christian fundamentalism for example, predates the global village and began with print media, radio, and television. Early in the twentieth century in the United States, a flood of books, magazines, and daily newspapers devoted increasing amounts of space to the rise of science and modern technology. These accounts coupled with Darwin's theory of evolution cast doubt on the Biblical account of creation. Darwin's theory of evolution was in some ways the separation point. Many Christians believed to the core of their beings that the Bible was sacred revealed truth and literally correct. If so, what was the point of Socratic dialogue or reasoned argument? What was there to discuss? The result under normal circumstances was that moderate Christians or those of other religions simply left the fundamentalist Christians alone on this issue. It has become an agreed "avoidance topic" in civil discourse today as a practical way to preserve social harmony.

That strategy works until a secular law or court order is enforced in a way that impinges on an individual's behavior, values, and beliefs to strike a serious blow to the group's core identity. For example, Christian fundamentalists for most of the twentieth century were politically neutral. Then, a series of court decisions, laws, and cultural shifts took place that led them to feel their core identity was under a violent attack from science as well as a new social movement often called "political correctness." Islamic fundamentalists feel the same way for similar reasons. Secular non-believers are then viewed as persecutors who wish only to destroy the very core of their beings. Historically, one triggering issue for Islam was Israel's connections with the technically superior West.

For Christians, the biggest issues of threat were legal abortion, legal gay rights, the banning of school prayer, and the lack of mention of Christmas in or on public buildings. The laws and court decisions were, however, equally "fundamental" to liberal secularists' ideas of social equality and "freedom of choice." Political liberals have a fundamentalist wing of their own, one they share in part with many scientists and has been called secular fundamentalism, or SF. For most SF scientists, the issue is the nature of the universe. They see that nature as mechanistic, deterministic, and evolutionary, and they put great emphasis on separation between religious and secular life. Overwhelmingly, they are atheist and either anti-religious or at least sympathize with the Marxian view that religion is the opiate of the masses. At one time, this was a mainstream view of science that emerged from Newtonian physics in the seventeenth century. The emergence of newer sciences, especially quantum physics, Big Bang cosmology, nonlinear chaos (or complexity) theory, and, from biology, the study of self-organization and autopoiesis all undercut the mechanistic view of the universe. The new sciences even explored a rational belief in some sort of creative force or higher power that influenced the course of evolution in light of a broader purpose for the universe. (See Appendix 1 for a more detailed discussion).

A host of various groups arose to promote the SF political agenda. Broadly speaking, Christian fundamentalists (CF) and secular fundamentalists (SF) have been at odds with one another on that agenda since about 1970. Neither side compromises on core identity issues. Autopoiesis strongly inhibits either side from doing so. Moreover, each side is absolutely right, given their respective core assumptions. The result is that one side (thesis) will subdue the other side (antithesis) or else both will disappear in some synthesis that creates a new set of core values and identities in a new generation of young people. Until the 2004 U.S. election, the SFs appeared to be dominating the political agenda but less so on the scientific front. However, a new synthesis may be emerging among the public and many scientists that rejects atheism but accepts secularism. This new secular synthesis remains in a formative stage with many unsettled specific issues.

Internationally, unless secularism breaks out in a big way in areas where Islam predominates, good faith negotiations are not likely to resolve the current conflict over globalization and the threat it implies for sacred beliefs and traditions. One side must defeat the other. If so, then how can we find much basis for optimism in a contest between two intransigent views?

If history is any guide, it seems more than likely that the secular synthesis will win because it embraces technology and the freedom to innovate. As earlier pointed out, over time high-tech always beats low-tech. The present anarchy and retribalization of the world is a temporary departure from a longer run, and perhaps, inexorable trend. The current resistance to globalization is simply a short delay leading toward a world ruled by secular democratic and high-tech free market states that largely work from the same page. These states are already highly interdependent. That fact reduces the danger of war among them and is different from the interdependence of the nineteenth century, because the major powers were then mostly imperial powers. Such powers have always fought each other. It is now commonplace to observe that democracies, despite their many differences, don't fight physical wars with each other. Democracies can and do come together to resolve important issues including issues of technology despite other differences. Canada and the United States have many differences, but there are no forts on either side of their three thousand-mile border. Many warrior states of the past shed that warrior ethic when they became democratic and embraced a market economy. The Scandinavian nations, Japan, Germany, and France are examples. Market competition seems to preempt aggressive warrior-like urges. The imperial impulse remains in force but becomes market-focused displacement activity as corporations expand overseas or become multinational organizations. A country no longer must invade its neighbors to acquire needed resources; it can export its goods and services to earn the money to buy what it needs in a global economy featuring significant free trade.

High-tech, free-market capitalism thus has the potential to create an orderly world without the need for a central autocratic authority to impose its will. Additionally, the market system can tolerate a good deal of cultural diversity. Still, secular, reasonably democratic government and reasonably free markets and free trade need to become a global standard or nearly so. Small enclaves to the contrary might exist, even protected as such. Given the increasingly integrated high-tech global human habitat, a stable human ecosystem seems to require a choice of standards. A village can't succeed with several different and conflicting standards, or violence will erupt.

Do we return to a reactionary despotism with central world control or do we push ahead with democracy and all that it implies? Modern technology has successively breached wide oceans, Chinese walls, Iron Curtains, and even outer space. Thus, geographical separation no longer seems an option to protect diametrically opposed points of view on how best to run an interdependent global human society. It simplifies into a contest between constitutional democracy and despotism. One will win, the other will lose. Circumstances will likely determine the winner. Today's circumstances favor democracy and a continuation of humanity's million-year-plus evolution as nature's only high-tech species. A human-imposed stop to new technology worked only for about five thousand years, and then not perfectly. All the ancient despotisms succumbed to their fatal temptation to become kleptocracies. Despotism has no internal system of reform, because by its nature it suppresses internal competition. Democracy depends more on internal competition. Democracy is more likely to reform and adjust to changing times and circumstances that are always in a state of flux, and markets have a long history of such adjustments. Sometimes markets fail, to be sure, but huge adjustments have been made in markets in the past two hundred years.

Yet democracy's strengths as seen by Islamic fundamentalists are deeply heretical, and in gross violation of Allah's plan as revealed by the Holy Koran. Moreover, most Islamic fundamentalists deny our evolutionary heritage as high-tech creatures driven to make and remake their own cultures in a process of creative destruction. Evolution is not God's plan, Islamic fundamentalists insist, evolution is the way of the infidel. Islam, like nearly all religions of doctrine, aims for stable equilibrium and blocking further change. Most such religions tend to look askance on change, on the punctuation side of the evolutionary process. But if God wanted only stability and equilibrium, we would never have had a Big Bang. Or if we had physical creation, the earth would still be simply the third (lifeless) rock from the sun. Or if, having created the sun and the earth, God created life, then chose to keep it stable, we would all still be bacteria.

If evolution does reflect divine guidance by whatever Creative Powers that be, surely humanity and its immediate hominid ancestors have been following this divine plan for about six million years. That guidance seems to have steered us in the direction of inventing our own technology in response to enormous and sometimes traumatic disturbances in the physical world. Yet even for those who insist, as a matter of faith, that evolution is blind chance—one enormous improbability stacked on top of a large number of other enormous improbabilities—then simple fate seems to have dictated humanity's high-tech destiny. Neither assumption would seem to suggest that we should suddenly call a halt. It is also possible, regardless of what we do, that oblivion will be our fate. Oblivion, after all, has been the fate of more than 99 percent of all species that ever emerged. Indeed, all life on earth is doomed according to the standard view of the evolution of our sun. According to Peter Ward and Donald Brownlee, conditions favoring animal life as we know it can exist for only about one billion of the estimated ten billion year estimated life span of our planet. About half of that billion years has been used up.[4] A half billion years from now life will begin to devolve back to microbes. The earth itself ends when the sun enters its red giant phase that engulfs the planet, but its oceans will have boiled away long before that as the sun heats up. If life on planet earth is to escape this heat death, humans must invent the technology by which it can escape. We would have to design and build a space ark. We might also have to reengineer our DNA to equip us for long voyages in space.

Between random fate and divine guidance lies yet another possibility. Perhaps life as a whole (Gaia?) has a collective subconscious that can foresee this fate. Gaia thus self-organizes an escape by evolving high-tech humans much as it evolved wings on birds so that they could fly. This is, of course, sheer speculation. Yet human innovations in technology may well serve a cosmic purpose we don't currently understand.

Unless some drastic and unpredictable event occurs (something we can never rule out), the high-tech democracies will in the end subdue Islamic fundamentalism. Autopoiesis could drive Islamic fundamentalist to fight to the bitter end. We have a recent history of resistance in defense of a way of life and its sacred values, as with Americans in the South during the Civil War, and the Nazis and Japanese in World War II. The Palestinians under the strong influence of Islamic fundamentalists show every sign of continuing the trend.

Autopoiesis, however, is not the exclusive province of traditionalists. America's despotic adversaries in every war it has fought were convinced that Americans were too soft, too materialistic, and too genetically diverse. America was an impure mishmash of mongrel races in a state that could not stand up to the pure and superior spirit of (the South, the Master Race, Japan, Islam, etc.). However, all these high-spirited despotisms were crushed by superior technology. Moreover, America's secular democracy has revealed a spirit of its own over the years. Americans tend to exert their competitive spirit most of the time in commercial competition, to the disgust of the traditionalists. But when attacked as at Fort Sumter, Pearl Harbor, or the World Trade Center, autopoiesis immediately shifted that competitive spirit from its focus on commerce to a focus on the military mode that mounted a spirited defense. This vigorous response often comes as a surprise to America's enemies. What is more, that spirit can and has survived numerous setbacks as in the early defeats of both the Civil War and World War II, whereas our adversaries expected Americans to retreat or surrender to those setbacks.

This "bump in the road" toward globalization however, will bring plenty of trauma and turmoil in its wake and—from the West's point of view—a share of setbacks and defeats. But the West is not going to abandon its technology, democracy, and commitment to the free market. After all, every alternative thus far advanced has been a dismal failure, and even internal critics of capitalism have no real alternative to offer other than to yell stop. Since the Turks' crushing defeat before Vienna in

1683, the high-tech West has prevailed in every major military contest with Islam.

The new element is non-nation-specific terrorism. But such terrorism does depend on safe bases somewhere. In the September 11 attack, Arab terrorists used America's openness against it. That places America in the dilemma of forcing us to curtail the very freedoms that propelled us to become a superpower. Those restrictions are not acceptable to most Americans except for short periods of crisis. Not wanting to be a theocratic police state itself, America's only reasonable alternative is to mount an aggressive defense of fighting terrorists anywhere they are. So far that is what we have done. That strategy is controversial to be sure, but not as controversial as a police state alternative.

The future, of course, is not clear. An impasse lasting for decades cannot be ruled out. America has much to learn in dealing with terrorists who sacrifice their lives to preserve their sacred traditions and ways of life. But we will learn and invent ever newer anti-terrorist technology and covert ways of using it. Islam will not vanish but more likely will come to terms with modern science and technology much as most Christians, Hindus, and Buddhists have. Perhaps Judaism, Christianity, and Islam will someday find some common ground as Muhammad always hoped. Meanwhile, Asia, once a bastion of tradition, has made rapid strides toward this new secular and democratic standard. Perhaps the pace of technology will slow down on its own, perhaps in the near future. Human technology has successively shifted from Stone Age tools powered by humans, to agriculture and domesticated animal power, to wind, hydraulic, and finally mechanical power, to electrical power and electronics, and now into biotechnology and nano-technology. What will come after biotechnology? One might speculate that scientists will learn how to generate room-temperature fusion or super-conductivity, how to control gravity or tap into the infinite zero-point energy field said to pervade all space. That should forever end energy scarcity. Forty years ago, no one predicted that by 2005, we could send almost infinite amounts of digitized data anywhere in the world almost instantaneously and at almost no cost. Still, that advance has had its own downside by facilitating the terrorism that arises in part from the backlash to technology by traditional societies. However, we have many instances of anticipating a suite of technologies that have yet to emerge, such as room-temperature fusion. By 1956, many Americans thought that every middle-class family would have its own heliport and helicopter by 1986. The 1968 movie *2001: A Space Odyssey* showed space travel and human dependence on intelligent computers by that date as the way of life. However, emergence of future of technology cannot be based on current information because innovation depends on too many nonlinear relationships. One small change can cascade into a wholly unexpected outcome or new emergent condition. Conversely, the rapid rate of change in human cultures could slow. Stasis would then replace frontiers. Such a transition could involve much trauma, depending on the state of our culture. Full employment in our current culture requires that we continue innovating in order to keep demand growing. Should such growth cease, prolonged unemployment could follow under today's ground rules, where nearly all labor is hired labor, unless current policy changes.

Japan's organic business model could well become a world standard because of its ability to ease the transition from growth to maturity by maintaining employment via reduced earning rather than layoffs. In the process, such firms remain efficient. The organic model allowed Japan to escape mass unemployment during the Great World Depression of 1929-33 and to recover quickly from the devastation of World War II after 1945. Since the mid 1980s, many western firms have moved toward the Japanese model. Some have replaced the mechanistic buyer–seller relationship with organic (family-like) relationships between employer and employee to improve internal cooperation and the quality of their output. Outsourcing, however, tends to go in the opposite direction.

Can we look forward to the day of equilibrium without endless punctuations of new technology and the many disruptions in life they bring? Perhaps. However, external disruptions can always strike. For instance, Earth may confront a new Ice Age or its reverse, global warming. An asteroid may hit, or a tectonic plate shift could induce a series of earthquakes and tsunamis, perhaps worse than the 9.0 quake off Sumatra in late 2004. Any such future event could disrupt any equilibrium we may have achieved. Another punctuation period would erupt with new frontiers of evolution, and the FROCA process would once again drive evolution.

Barring cataclysm, the real question is whether we will or will not reverse, or at least slow down, the growth of the rapidly mounting and dangerous human impact on the global ecosystems. This impact is now estimated at one hundred thousand times as large as when humans turned to agriculture, and most of that increase has occurred in the last sixty years. Our current rate of increase is simply not sustainable without some kind of a major correction. A central world government with the legitimate authority and power to manage the global ecosystem as a whole does not appear possible any time soon. The inability of the United Nations to manage any crisis where conflicting national interests are at stake illustrates this.

But could our own innovations in technology, the major force that led to the dangerous degree of human impact on the global environment, actually begin to reverse that rate? There is one hopeful sign that it could. We talk now of a post-industrial world where heavily polluting heavy industry no longer dominates the economy as it has historically. For example, Pittsburgh, Pennsylvania, was once the iron and steel capital of the world, but today it manufactures no steel at all. More broadly, in the past thirty years or so the work force has shifted from processing raw materials to processing information. The energy thus required to produce a unit of gross national product has fallen, by as much as 50 percent. That trend, at least, is well under way.

Still, the future is unpredictable and the increase in human impact now appears to be driven more by improved standards of living in non-western nations such as China who aspire to western standards of living. In such rapidly growing nations, western living standards in good part still depend on older polluting technologies. China, for example, now produces more steel than any other nation and has a long history of inflicting enormous environmental damage on itself. Its cities are among the worst polluted in the world. But signs of hope are evident such as China banning all logging in its remaining forests. Then too, some new technologies, such as cell phones, don't require nearly the infrastructure the earlier land-line telephone technology required. That means much less wood for poles and almost no copper for wire.

It is not realistic to assume that the world will escape serious consequences of so increasing the adverse human impact on the environment. The exploitation of new frontiers by all life, particularly human life, leads to overexploitation followed by a crash. Humans almost certainly have overexploited the opportunities our innovations in technology have created for us, that overexploitation now takes place on a global, not only on a local scale. Thus, history suggests that we can expect a crash to follow, but we can only guess at when. But let us remember that these crashes have always played a vital role in evolution itself—no punctuation phase, no evolution.

The human species as a whole has always rebounded from its crashes big and little, and presumably will do so again after any future crash. Where human life is concerned, we cannot know what long-run outcomes will be. We can say that human life, for all its mistakes, has demonstrated infinite potential for variety, change, and adaptation.

APPENDIX 1

SCIENCE AND EVOLUTION

The idea that human culture evolves much as human anatomy does has gone through several cycles of acceptance and rejection within the scientific community. Those, such as Franz Boas, who strongly rejected the idea of cultural evolution seemed to fear that it put a stamp of approval on the blatant ethnocentrism then common in western culture. In more recent years, biologists who study evolution accept cultural evolution as a fact.[1] For one thing, many distinct cultures have evolved through similar stages of increasing complexity in the last ten thousand years, and all evolved from the same Neolithic hunter–gatherer base culture. Some differences that have emerged can be easily explained by differences in climate, geography, and other environmental factors such as the presence or absence of animals available for domestication. Still, without technology humans could never have migrated to the colder climes and survived.

A more enduring controversy concerns the presence or absence of a "plan" or some form of "guidance" in the course of evolution, not just of life, but of the whole cosmos. Is the universe all purely random chance modified by physical necessity as some Darwinists such as Jacques Monad, Richard Dawkins, and Daniel Dennett insist? The question of "critical complexity" in the emergence of traits that clearly provide a survival advantage raise certain questions. Michael J. Behe argues that critical complexity implies such huge degrees of improbability as to negate the logic of pure chance.[2] Behe argues that critical complexity arises when, for instance, five separate mutations must occur in a definite sequence before the trait emerges that provided survival advantage. How could natural selection select the components before the whole trait emerged to provide an advantage? If each component mutation provided no extra fitness by itself, then why was it selected? Why would it wait patiently for the other components to arrive by random chance? Behe argued some kind of divine guidance is at work, interacting with randomness, to explain critical complexity. Behe is not a "Biblical creationist." He accepts evolution as proven beyond doubt. But he also argues that random chance by itself cannot explain such facts of evolution as critical complexity.

After I read Behe, I looked up the reviews written about his book on Amazon.com. I found 139 reviews listed at the time. A few were reasoned dissents that suggest it was simply not obvious to us in what way the intermediate mutations did provide enough fitness to survive. Far more dissenting reviews were apoplectic and personally abusive. Many were sheer rants that did not even try to refute his arguments. Some reviewers were clearly outraged that a Ph.D. biochemist would presume to cast doubt on Darwin's "Revealed Truth."

This experience opened my eyes to the existence of secular fundamentalism. Behe's critics were as intransigent as Christian, Jewish, or Islamic fundamentalists. It seems that randomness is a hot button issue to secular fundamentalists. Behe was looking at the complex detail of cellular life not suspected until long after Darwin wrote. No one then had any idea that a single cell had up to fifty thousand separate component parts. Clearly, such complexity does not fit comfortably into Darwin's model built purely on the natural selection for fitness of random mutations. Whether or not Behe is correct, he makes a perfectly reasonable case within the usual parameters of science. Only those who, as a matter of faith, rule out the existence of some nonmaterial force or influence in the process of evolution could claim otherwise.

Such a "ruling out" was mainstream scientific thought for many decades after Darwin first wrote. Those were the formative years of a strong and solid belief in a purposeless, random, and mechanistic universe. At the time, that view was simply a best judgment based on the best scientific evidence then at hand. To a secular fundamentalist, however, that belief has acquired the status of "revealed truth" that in principle cannot be refuted.

Today, however, *all* the assumptions supporting mechanistic determinism have been disproved or at least undercut by newer and more sophisticated sciences. These new sciences include quantum physics, Big Bang cosmology, chaos (or complexity) theory, and theories of self-organization and autopoiesis, together with a much better and more detailed view of the complexity of the living cell that Behe discussed. These newer sciences do not, repeat, do not, *demand* a belief in some divine presence, a universal mind, a cosmic consciousness, or just plain God. Still, they certainly open the door to look at such beliefs. Many well-regarded scientists have chosen to walk through that door. My FROCA analysis, however, in no way depends on such a belief. Yet it is compatible with one. To me, the "purpose" if any, of evolution simply remains an open question.

I say this as one who was, for most of my adult life, a secular fundamentalist, of sorts. I had allowed for the possibility of some kind of creative force that science had not identified that brought the universe into being. I developed my agnosticism in college. In those days, science teachers assured me the universe was in a steady state. I was further assured by that great French mathematician, Pierre Laplace, that the future was embedded in the present. The future would flow inexorably from the present in a linear fashion and in accordance with Newton's inverse square law of gravity, and his three laws of motion. My science teachers assured me that the only "real" reality was physical reality. They went on to note that a scientist could conduct experiments objectively without participating in the experiment itself. Free will was an illusion, I concluded, perhaps necessary to human society, but an illusion nevertheless. I also believed the news stories of that day that biochemists were about to create life in a test tube since they had successfully created the necessary amino acids that way. I was taught that a living cell was an undifferentiated mass of protoplasm. I became one with Gilbert Ryle who dismissed the Cartesian idea of a mind apart from a brain as "the ghost in the machine."[3]

Then, beginning in the early seventies, I learned about the Big Bang theory of creation and of the cosmic background radiation that seemed to confirm it—so much, it seemed, for a steady state universe. Then I learned about the Heisenberg Principle of Uncertainty: to fix the position of a particle, one must necessarily disturb its velocity or to fix the velocity, one must disturb its position—so much for the purely objective experiment with observer-free outcomes. I then learned of the wave/particle duality of matter and how an observer could collapse a wave state to a particle state and back again as often as one liked. I learned that the wave contained no matter or energy. It contained only some probabilities, one of which an observer made real by collapsing the wave function to create matter and energy. So much for the idea that the only reality was physical (matter and energy). Paul Davies and John Gribbin noted that "Ryle was right to be sarcastic, not because there is no ghost, but because there is no machine!" Suddenly, it seemed possible that some sort of universal mind had to collapse the first wave function to create a Big Bang. Then I learned about nonlocality and was shocked to learn of the Pauli Exclusion Principle where electrons jumped from one state to the next while in a shell around the nucleus of an atom, and without traveling in between. To my further dismay I learned that each cell in my body contained my complete DNA and as many as fifty thousand moving parts—so much for protoplasm, whatever that was. Then I came across Chaos theory with its butterfly effect based on nonlinear turbulence and where one equation could produce several outcomes and how the tiniest change in initial conditions, via positive feedback loops, could cascade into vast changes in outcomes.

By about 1985, I could no longer sustain the worldview of a mechanistic, deterministic, and largely pointless universe I'd picked up in college. These strange new sciences had shot down all the major assumptions that sustained the Newtonian worldview. I was, of course, aware that many scientists remained wed to that worldview and concocted several ad hoc untestable hypotheses to defend it. My favorite, based on not a single shred of actual evidence, holds that every time a wave function collapses into a particle of actual matter, that same particle emerges at all other potential positions at once, but each site is in a different or parallel universe.[4] This defense, it was said, retained the math of the standard "Copenhagen Interpretation" of quantum physics in tact. The "many worlds" hypothesis allowed the atheists who were forced to use that math in their daily work to deny the logical nonmechanistic implications of the wave particle duality. That implication holds that the universe contains, aside from waves, particles, fields of force, some sort of a universal mind or cosmic consciousness that collapsed the nonmaterial meta wave function into the Big Bang then into particles of real matter in a field of force. I now see the many worlds defense (including its variant the "summed over histories") more as an example of autopoiesis at work, not as valid science. Here, autopoiesis is working to preserve the core identity of secular fundamentalists for whom mechanistic determinism is an article of faith, not just a question of philosophy. Power to define Truth is also at stake. Take away mechanistic determinism and science loses its role as the final arbiter of Truth.

To me, the many-worlds defense is the equivalent of the Biblical literalists who dismiss the fossil evidence of evolution with the untestable hypothesis that God salted the fossil sites simply to test our faith. The issue is the same in either case: preservation of power. When science finally proved in the seventeenth century that the earth was no longer the center of the universe, and later that it was vastly older than six thousand years, and finally that all the species had evolved rather than having been created in place fully formed, theologians lost their place as the final arbiters of Truth.

Physicists soon took over that powerful position based on their claim that the only reality was physical reality. Thus, reality was something only scientists were in a position to understand. Christians (and later Muslims in a different way) reacted by reverting to a fundamentalist strategy. Both groups of fundamentalists created for themselves little worlds of faith where they remain final arbiters of Truth. Now secular fundamentalists are doing much the same thing, defending their "rightful" position as final arbiters of the Truth. All three groups use intransigence to preclude any meaningful dialogue with those who dispute their claims. This is not an "irrational" stand so much as a defense of a core identity, and it is thus quite natural. We humans almost never apply "reason" or "critical thinking" to undercut our own vital core identities or the beliefs that sustain them. We have strong innate inhibitions from doing that sort of thing unless those beliefs propel us into some kind of crash. It is only then that most people will go through the agony of a major reevaluation and create a post-crash core identity.

Again, the question of whether or not there exists a broader purpose or "goal" of evolution, divine or otherwise, remains to me very much an open scientific question.

APPENDIX 2

AUTOPOIESIS: TOWARD AN INTEGRATED THEORY OF HUMAN BEHAVIOR

Western culture's well-established approach to higher education was heavily influenced by the rise of modern science in the seventeenth and eighteenth centuries. As it became clear that the natural world was more complex than we had thought, academicians began to place the study of human behavior into separate compartments. Compartmentalization made it easier to focus on a specialty. But, this sort of reductionism was also thought to make it easier to learn about the world around us. Only in part were the academicians correct. If we understand the broad nature of a whole, reducing it to its component parts is the best way to come to know how the parts work and fit together. If, however, we don't understand the whole, reductionism (an analysis of the parts) will not much help. (Suppose an engineer in 1800 somehow acquired a modern desktop computer. If so, I doubt that he would ever deduce its purpose simply by pulling it apart.)

Thus, to this day there is no integrated general theory of our human behavior. When academics began to divide human behavior into separate disciplines they created a category called the "social sciences." This act inhibited the development of an integrated theory. The separate social sciences emerged in a sort of random fashion. My field, economics, was the first to emerge as a separate discipline. Economics is the study of human behavior in the marketplace. It excludes other categories of behavior that can influence economic behavior. Soon after economics appeared, the field of sociology emerged to study the structure of society and behavior of people within it (usually within some culture). Cultural anthropology followed soon after to study behavior in different cultures. The focus at first was on more primitive cultures. Psychology emerged about the same time to study the behavior of individuals, beginning mainly with Euro-Americans. Political science soon emerged to study political behavior. At its root, politics is the quest for power and then the application of that power. Power, of course, is something philosophers at least since Plato had long studied. Yet with all these separate disciplines we still have no theory that integrates them into a general theory. All the social sciences offer useful insights and valid partial theories but no general theory. In all fairness, only since the 1970s has science itself advanced far enough to let us generalize about human behavior in a way that firmly roots that behavior in our biology.

Darwin started out in the right direction except for the fact that he had a personal bias in favor of "uniformitarianism." He acquired uniformitarianism from Charles Lyle's theory of geology that ruled out "catastrophes" as a way to explain the evolution of the physical earth. In the view of some critics, uniformitarianism (much like the many worlds hypothesis) was based not on evidence but on an attempt to distance science from the Bible's account of geology found in Genesis. Even so, Darwin's theory of natural selection of random mutations did unite the then-separate sciences of botany and zoology into the common science of biology that we know today. Uniformitarianism, however, led Darwin to propose that natural selection was a continuous process unhindered by the kind of catastrophes and disruptions that we now know are routine events on the time scale of evolution. Moreover, his theory of continuous change never did fit the fossil evidence (which was, again to be fair, scant by today's standards).

A major difficulty with Darwin's continuous change hypothesis is that some primate species of monkeys and apes have remained more or less constant for millions of years. Yet, one branch of the apes, the hominids, evolved rapidly as earlier discussed at length. The hominids evolved, branched out, and one of those branches in turn evolved into early humans such as *Homo erectus* and later, by stages, evolved into a single species of modern human.

Four New Sciences

When Niles Eldredge and Steven J. Gould, the paleontologists who finally fit the fossil evidence to evolution, came on the scene, they made that fossil evidence fit by dropping Darwin's hypothesis of continuous change. Their theory of punctuated equilibrium was a crucial breakthrough. It was thus the first of the new (or newly modified) sciences that make a general theory of human behavior possible. In addition, ecologists such as Tim Flannery who studied frontier ecology noted the progression of steps in a frontier (that I called FROCA). The FROCA steps specify different kinds of behavior found on a new frontier following a major disruption to the local ecosystem. These insights explain why it is that most actual change happens on punctuated frontiers and not in stasis or stable equilibrium. Thus, we have noted again and again how human behavior systematically differs depending on whether people find themselves living within a frontier of change or in a culture bound in stasis by its cultural traditions. Traditions can be thought of as social software that aims to limit change and keep things stable. Results trump tradition and rules on the frontier; rules and tradition trump results for a culture in stasis.

Chaos (or Complexity) Theory also came into public awareness in the 1970s and 1980s with James Gleick's book, *Chaos: Making a New Science*. This new theory improved our understanding of self-organization and of how order can suddenly crash into disorder, and alternatively, order can suddenly emerge from chaos. These shifts occur both with animate and inanimate matter. Chaos theory has also improved our understanding of the contingent nature of the universe, and helped end the once widely believed philosophy of mechanistic determinism. Such determinism assumes a mathematically linear relationship, one where small changes in initial conditions produce comparably small changes in later outcomes. Relationships that are linear break down in the turbulence we find throughout nature on both a physical and social level. Given turbulence, nonlinear conditions prevail, where small changes in initial conditions can produce enormous changes in later outcomes. A nonlinear equation can often produce a set of equally likely but quite different outcomes with no way to tell which will occur. Gleick recounts how this was first understood at MIT by weather forecasters who coined the whimsical term "butterfly effect" to suggest the implications. If a butterfly were to flap its wings in just the right way at the right time, it could set up a positive feedback loop that cascades into a tornado as much as five thousand miles away six months later. Unless we can track all the butterflies and all else that moves, we can't make accurate weather forecasts for more than a couple of days. If we could track all the butterflies and other moving creatures and objects, the huge volume of instrumentation would then hopelessly distort the very weather we wanted to forecast. Human observation, in other words, can and often does influence behavior in the macro world and not only in the micro world of quantum physics and its Heisenberg Principle of Uncertainty: to observe the velocity of a particle distorts its position and to observe position distorts the velocity. We can never be certain of both the position and velocity of a particle as a result. By the same principle, we can't be sure how a person will react to a given stimulus if we observe the reaction. The reason is that the observation itself can and often does change the reaction. For example, people with radar detectors usually slow down if they detect radar surveillance. Shoplifters may not steal when they see an aisle monitored by closed-circuit cameras.

Another modification of biology also appeared in the 1970s, namely the concept of autopoiesis, or self-making. Humberto Maturana and the late Francisco Varela coined the term to account for the rigorous way in which life strives to remain self-consistent in response to the continuous flux and change in nearly any living environment. Thus, each individual cell has evolved the capacity to repair damage to its own DNA, each organ can also repair itself up to a point, and the immune system of the body can spontaneously respond to toxic invaders, whether they are poisons, viruses, or bacteria. Thus, to a great extent, all life is in good part self-organized. In a sense, Maturana and Varela fleshed out the "law of self-preservation" that has long been a part of common wisdom. Combined with Chaos Theory, autopoiesis gives us a much better understanding of human behavior.

Autopoiesis and Identity

Autopoiesis allows us to make sense of some otherwise puzzling aspects of behavior that involves self-image or self-identity. For example, if the law of self-preservation is correct, then how can we explain such contradictions as the Japanese kamikaze pilots of World War II or more recently the suicide pilots that flew into the World Trade Center in 2001, or many recent instances of suicide bombers in Palestine or Iraq? Such behavior seems more an act of destroying one's identity, not protecting it. But we realize that each human has, in addition to a personal identity, a higher identity as well. Indeed, we all have multiple identities but they are not all of equal value to us. Our personal identity is spread out over the many different kinds of relationships we form and can elicit different behaviors depending on which side of a relationship we are on, for example, parent–child, teacher–student, doctor–patient, husband–wife, brother–sister, employer–employee, master–servant. As social animals, we humans usually identify with a group or other entity beyond or above ourselves. That higher identity will often become our core identity, one that takes precedence over our other identities if they conflict. In many cases, multiple identities fit comfortably together without conflict. For example, an individual can be a patriotic citizen, member of a particular family, clan, or tribe, an employee of a company or other organization, and a member of a religious group all at once, with little conflict between them. At any one time, any of these identities may be a core identity, and the core may shift from time to time depending on circumstances.

According to Samuel P. Huntington, when civilizations with different core religions are clashing, the religion of each respective civilization soon becomes a major core identity for most people in that civilization.[1] That can happen even when religion was not a major force before the clash. Huntington gives a compelling example from Yugoslavia before and after its breakup following the death of Marshall Tito. Yugoslavia contained adherents from three religions, each of which is at the core of a major civilization. The religions were Catholicism (Slovenia and Croatia), Orthodox Christianity (Serbia), and Islam (Bosnia, Kosovo, and nearby Albania). Before the breakup, the adherents of the three religions lived in harmony with each other, often intermarried, worked together, and lived together with little tension. Yugoslavia under Marshall Tito was a communist state and was officially atheistic. Thus, it is perhaps not surprising that the adherents of all three religions were noted for their religious laxity, with low attendance at the respective religious services. To quote Huntington, "Historically, communal identities in Bosnia had been strong: Serbs, Croats, and Muslims lived peacefully together as neighbors: intergroup marriages were common; religious identifications were weak. Muslims were Bosnians who did not go to the mosque, Croats were Bosnians who did not go to the cathedral, and Serbs were Bosnians who did not go the Orthodox church." But, after the breakup of Yugoslavia, Huntington notes, "...these casual religious identities assumed new relevance, and once fighting began they intensified. Multiculturalism evaporated and each group increasingly identified with its broader cultural community and defined itself in religious terms."[2]

We see in Huntington's example a shift of core identity in response to a perceived threat, a shift from a casual to an intense core identity. Like any other personal "possession," we may take an identity for granted until we perceive that someone or something threatens to take it from us or degrade it. Then we can become very protective. The communist state had rejected all religion, so none were singled out. Thus, people did not have to choose between them. In an aggressively secular state, it did not serve to make one's religion the center of one's identity. But once Yugoslavia's central control began to erode, the ethnic independence became politically possible for the separate ethnic entities such as Croatia, Bosnia, Slovenia, and Serbia. Long-standing animosities separating the different religions surfaced. At that point, local religious identities quickly sprang up. The principal at work here seems to be as follows: Autopoiesis impels all life to protect its identity. For humans, this defense works both for individuals and collectively for culturally connected (or bonded) groups of individuals, even when such a defense can seem irrational to outsiders. In his latest book, *Collapse,* Jared Diamond describes an interesting example for the Greenland Norse in the fifteenth century. They were so bonded to their identity as European Christians and the way of life they felt went with it, the Norse refused to adopt the practices of the pagan Inuit that would have let them survive. Thus, the Norse starved to death in the Little Ice Age that had made their way of life virtually suicidal as the temperatures continued to drop. A similar determination to continue to practice settled ways probably led to the demise of the Easter Islanders as well.[3]

Threats to the relative status of one's identity can elicit autopoietic defensive responses. Such defensive responses again apply to both individuals and socially connected groups of individuals. If such threats persist, identity-defense can become intransigent, or fundamentalist. Such defensive behavior has persisted as long as has recorded history and in a wide variety of circumstances across cultures. Many scholars of Islam such as Huntington, Armstrong, Lewis, and others have noted the strong sense of humiliation many Muslims feel about the relative collapse of power and status of Islam from its peak period from about 800 CE to 1400 CE. Western Europe was then an underdeveloped civilization. After 1400, however, western Europe began the rapid techno-driven recovery outlined in Chapters 12 and 13. Not wanting to copy the practice of Europeans, Islam ignored the new technology coming out of Europe. The Turks, for example, knew of the printing press for about three hundred years before adopting it. Armstrong has noted that some interpretations of certain of its passages suggest that the holy Koran looks upon human innovations as evil. Like the Norse who refused to copy the pagan Inuit and eat the plentiful fish around Greenland, Islam long resisted adopting western innovations to the point where Europe became the world's dominant culture. Islam, in terms of the modern world, became a backwater. Today, Islam does indeed adopt western military technology. Yet, Islamic fundamentalists who seem to be in charge still resist departing from their holy scripture by adopting the "cultural software" of constitutional democracy that makes sustained innovation possible. The fundamentalists of Islam are convinced that Allah wants the entire world to be governed according to Koranic law, the Shariah. A major strategic departure from that goal puts the core identity of Islamic fundamentalists at deep discount.

In an earlier book, I showed how autopoiesis explains the worldwide phenomenon of bureaucratic defensiveness. An obsession with rules, a tendency to overstaff, to over-layer, and to resist internal innovation, often in the face of serious threat, all follow. Yet, autopoiesis equally well explains, when combined with punctuated equilibrium, why start-up firms are so much more innovative, flexible, and more inclined to ignore rules that obstruct birthing a successful new enterprise. Autopoiesis also explains why almost any large organization in almost any culture will be passionate about the details of protocol. Who must defer to whom, who sits where at the table and who at the "head table," who sits on the right or the left of the chief, in what order of precedence is one to be introduced at Court, who gets the corner office, and so on, all define relative differences in a person's status within a group. Autopoiesis drives all social animals to pay close attention to social status and protect that status. Such behaviors cut across all the social sciences as applied to any culture.

This appendix, however, aims to outline some of the parameters of an integrated general theory of human behavior, not a completed theory. I think the evidence is persuasive that the four concepts from the new sciences discussed earlier make it possible to construct such a theory. I hold that biology's concept of autopoiesis is the vital core of an integrated theory, but one strongly buttressed by punctuated equilibrium from paleontologists, the frontier effect as outlined by ecologists, and finally by Chaos or Complexity theory from physics. All of these concepts offer vital insights on the self-organization that governs so much human behavior. Combined, these four new concepts point toward an integrated theory of behavior connected to biology. It offers the social sciences a common base that they can then flesh out with their own special and vital insights with far fewer contradictions between them.

NOTES

Introduction

1. Michael J. Benton, *When Life Nearly Died: The Greatest Mass Extinction of All Time* (London: Thames and Hudson, 2003), 1–17. Benton reviews all the major extinctions since the origin of the earth.
2. *The Pirates of Silicon Valley* (AV, Turner Home Video, 2000). This movie gives a dramatized but roughly accurate account of the formation of both Microsoft and Apple Computer focused on rivalry between Bill Gates and Steve Jobs.
3. Charles Darwin, *The Origin of the Species* (New York: Norton, 1975). For a more recent but faithful reflection of Darwin's original theory see Richard Dawkins, *The Selfish Gene* (New York: Oxford University Press, 1979).
4. Benton, *When Life Nearly Died*, 56–70. Benton discusses uniformitarianism at length.
5. Niles Eldredge, *The Pattern of Evolution* (New York: W.H. Freeman, 1999). See also his *Time Frame: The Rethinking of Darwinian Evolution and the Theory of Punctuated Equilibria (New York: Simon and Schuster, 1985).* Steven J. Gould gives a comprehensive account of his life-work in *The Structure of Evolutionary Theory* (Cambridge, MA: Harvard University Press, 2002).
6. Thomas L. Friedman, *The Lexus and the Olive Tree* (New York: Anchor Books, 2000), 29-32. See also, Benjamin R. Barber, *Jihad and McWorld: How Globalism and Tribalism are Reshaping the World* (New York: Ballantine Books, 1996).
7. Samuel P. Huntington, *The Clash of Civilizations and the Remaking of the World Order* (New York: Simon and Schuster, 2003), 109.
8. Bernard Lewis, *The Crisis of Islam: Holy War and Unholy Terror* (New York: Random House, 2003), 53. In his introduction, Lewis quotes Osama bin Ladin's complaints against the USA at length. Part of his solution is for the United States to convert to Islam.
9. Information under www.silk-road.com/artl/timur.shtml provides a concise summary of Tamerlane's career.
10. William McDonald Wallace, *Postmodern Management: The Emerging*

Partnership Between Employees and Stockholders (Westport, CT: Quorum Books, 1998). Chapter 7 discusses the Great Depression at length.

11. Starhawk (pseudonym), *Webs of Power: Notes from the Global Uprising* (Gabriola Island, BC: New Society Publishers, 2002). 1–5. Starhawk was a principle organizer of the November 1999 WTO riots in Seattle and she stresses the use of the Internet as a factor in her success.
12. Humberto Maturana and Francisco Varela, *Autopoiesis and Cognition: The Realization of Learning* (London: Reidl, 1980).

Chapter 1. The FROCA Process of Cultural Evolution

1. Paul Ehrlich, *Human Natures: Genes, Cultures, and the Human Prospect* (London: Penguin Books, 2000).
2. Tim Flannery, *The Eternal Frontier: An Ecological History of North America and Its Peoples* (New York: Atlantic Monthly Press, 2001).
3. Jared Diamond, *Guns, Germs, and Steel: The Fates of Human Societies* (New York: Norton, 1999).
4. Ehrlich, *Human Natures,* 155–6.
5. Ibid.
6. Rondo Cameron and Larry Neal, *A Concise Economic History of the World: From Neolithic Times to the Present* (New York: Oxford University Press, 2003). On page 31 the authors state, "Almost all the major elements of technology that served ancient civilizations—domesticated plants and animals, textiles, pottery, metallurgy, monumental architecture, the wheel, sailing ships, and so on—had been invented or discovered before the dawn of recorded history."
7. Ibid., 42.
8. Flannery, *Eternal Frontier,* 238 and on, with my interpretation.
9. Ibid.
10. Much of the difference between the Democratic and Republican parties involves this debate between the imperatives of growth and those of economic justice.
11. Harold Evans, *They Made America: From the Steam Engine to the Search Engine: Two Centuries of Innovators* (New York: Little, Brown, 2004).
12. *The New York Review of Books*. November, 2004. From the outbreak of the Iraq War in early 2003 to the general election of November 2004, The New York Review of Books published so many articles to this effect that the author lost count.

Chapter 2. Climatic Crisis and the Risen Ape

1. Diamond, *The Third Chimpanzee: The Evolution and Future of the Human Animal* (New York: HarperCollins, 1992).
2. Ehrlich, *Human Natures*.
3. Richard Klein and Edgar Blake, *The Dawn of Human Culture: A Bold New Theory on What Sparked the "Big Bang" of Human Consciousness* (New York: Wiley and Sons, 2002). My account of the shift from the apes to hominids is taken from the above three texts. Also, see page 78.

4. Klein and Edgar, 66.
5. Ian Tattersall, *Becoming Human: Evolution and Human Uniqueness* (New York: Harcourt, 1998).

Chapter 3. The Ice Ages and a Bigger Brain

1. Flannery, *Eternal Frontier*, 136.
2. William H. Calvin, *A Brain for All Seasons: Human Evolution and Abrupt Climate Change* (Chicago: The University of Chicago Press, 2002), 249.
3. Ibid.
4. Ehrlich, *Human Natures*, 105.
5. Ibid., 107.
6. Ibid., 108.
7. Diamond, *Collapse: How Societies Choose to Fail or Succeed* (New York: Viking, 2005). Chapters 6, 7, and 8 lay out the history of the Greenland Norse.

Chapter 4. Techno-Culture Takes Over

1. Flannery, *Eternal Frontier.*
2. Ibid., 181.
3. Ibid., 188.
4. Ibid., 190.
5. Ibid., 191.
6. Ibid., 183.
7. Ehrlich, *Human Natures*.
8. Ibid., 243.
9. Ibid., 244.
10. Jonathan Kingdon, *Self-Made Man: Human Evolution from Eden to Extinction?* (New York: Wiley and Sons, 1993), 86–87. As his title suggests, Kingdon also believes that human inventions of technology largely account for our cultural evolution.
11. Diamond, *The Third Chimpanzee*, 198.

Chapter 5. Farming as a New Frontier

1. Cameron and Neal, *A Concise Economic History of the World*, 24.
2. Ibid.
3. Diamond, *Guns, Germs, and Steel*, 88.
4. Ibid., 91.
5. Diamond, *The Third Chimpanzee*, 262–72.
6. Diamond, *Guns, Germs, and Steel*, 8. Diamond lays out a flow diagram of his thesis.

Chapter 6. Release and Exploitation: A Population Explosion

1. Karl Wittfogel, *Oriental Despotism* (New York: Yale University Press, 1963). This was the main thesis of his book.
2. Diamond, *Guns, Germs, and Steel*, 283–4.
3. Ibid., 265–6, 277.

Chapter 7. Social Crash as Intra-Tribal Anarchy

1. Isabel Hilton, "The Pashtun Code," *New Yorker*, December 3, 2003.
2. Diamond, *Guns, Germs, and Steel*, 263–78.
3. Ibid.
4. D. J. Gray, *William Wallace: The Man Who Was Braveheart* (New York: Barnes and Noble Books, 1991).
5. Tony Judt, "Why They Hate Us," review of Thierry Meyssan's book, *September 11, 2001: The Big Lie* in *The New York Review of Books*, May 1, 2003.

Chapter 8. Adaptation through Despotic Civilization

1. Thomas Hobbes, *The Leviathan* (New York: Penguin Classics, 1982).
2. Diamond, *Guns, Germs, and Steel*, chap. 14.
3. Ibid., 286.

Chapter 9. Equilibrium through Repressed Technology

1. Cameron and Neal, *A Concise Economic History.*
2. Diamond, *Guns, Germs, and Steel*, 257.
3. Ibid.
4. Ibid, 259.
5. Cameron and Neal, *A Concise Economic History*, 85.
6. A shortage of funds may also have played a part in turning inward.

Chapter 10. Empires Evolve to Quell Inter-Tribal Warfare

1. Cameron and Neal, *A Concise Economic History*, 26.
2. Karen Armstrong, *Islam: A Short History* (New York: Random House, 2000), 8.

Chapter 11. Barbaric Disruption: The Fall of Rome

1. Michael Grant, *The History of Rome* (New York: Scribner, 1978), 255–6.
2. Cameron and Neal, *A Concise Economic History*, 96.
3. Armstrong, *Islam*, 73.

Chapter 12. Feudalism Opens New Frontiers of Technology

1. Jacques Barzun, *From Dawn to Decadence* (New York: HarperCollins, 2000), 171.
2. Ibid., 75.
3. James Burke and Robert Ornstein, *The Axemakers Gift: Technology's Capture and Control of Our Minds and Culture* (New York: Tarcher/Putnam, 1995), 140–1.
4. Barzun, *Dawn to Decadence*, 4.
5. Charles Seife, *Alpha and Omega: The Search for the Beginning and End of the Universe* (New York: Penguin Books, 2003).

Chapter 13. Europe Revives to Exploit Colonial Frontiers

1. Diamond, *Guns, Germs, and Steel*, 74.
2. Max Weber. *The Protestant Ethic and the Spirit of Capitalism*, trans. Talcott Parsons (New York: Scribner, 1958).

3. Barzun, *Dawn to Decadence,* 106.
4. Wallace, *Postmodern Management,* 24–30.
2. Flannery, *Eternal Frontier,* chaps. 21 and 27.
3. Alexis de Tocqueville, *Democracy in America* (Washington Square Press, 1964).
7. Debra L. Spar, *Ruling the Waves: Cycles of Discovery, Chaos, and Wealth From the Compass to the Internet* (New York: Harcourt, 2001).
8. Ibid.

Chapter 14. Democracy Challenges Despotism

1. Frederick Jackson Turner, *The Frontier in American History* (New York: Henry Holt, 1893). Reprint, 1947.
2. Spar, *Ruling the Waves.*

Chapter 15. New Technologies Open Industrial Frontiers

1. Wallace, *Postmodern Management,* 30.
2. Ibid., 29–32.
3. Ibid.
4. Karl Polanyi, *The Great Transformation* (Boston: Beacon Press, 1957).
5. Spar, *Ruling the Waves,* 78.
6. Thomas L. Friedman, *The Lexus and the Olive Tree* (New York: Anchor Books, 2000).

Chapter 16. America's New Frontier

1. Flannery, *Eternal Frontier,* 238–6.
2. Ibid.
3. Ibid.
4. Armstrong, *Islam: A Short History,* 102.
5. Spar, *Ruling the Waves,* 18–22.
6. Turner, *The Frontier in American History*

Chapter 17. Clashing Founder Effects and the Civil War

1. Albert Fishlow, *American Railroads and the Transformation of the Ante-Bellum Economy* (Cambridge, MA: Harvard University Press, 1965), 509–10.
2. de Tocqueville, *Democracy in America*
3. Jay Winik, *April 1965: The Month That Saved America* (New York: HarperCollins, 2002), 193–4.

Chapter 18. The High-Tech Frontier Takes Over

1. Michael Cox and Richard Alm, "Time Well Spent: The Declining Real Cost of Living in America," *1997 Annual Report, Federal Reserve Bank of Dallas,* 2–24.
2. Wallace, "Supply and Demand Interaction Model," *The Boeing Co.,* 1978.
3. Wallace, "Forecast Document: TSR 747," *The Boeing Co.,* 1961.
4. Wallace, *Postmodern Management,* chap. 7.

Chapter 19. America as Sole Superpower

1. Diamond, 104–13.
2. Frederick Morton, *Thunder at Twilight: 1913–1914* (New York: Scribner, 1989).
3. Diamond, *Collapse; How Societies Choose to Fail or Succeed.* Chapters 6–8 provide a complete account.

Chapter 20. Globalization and Its Discontents

1. See, for example, Armstrong (*Islam: A Short History*) or Lewis (*Crisis of Islam*).
2. Armstrong, *Islam: A Short History*, 23.
3. Ibid.
4. Calvin, *A Brain for all Seasons*, 225–9.
5. Huntington, *The Clash of Civilizations and the Remaking* of *World Order.*

Chapter 21. Resurgent High-Tech Tribalism and a Crash of the Western Alliance

1. William L. Shier, *The Rise and Fall of the Third Reich* (New York: Simon and Schuster, 1959). Reprint, New York: Touchstone, 1981. Shier stressed this point in much of the first half of his book.
2. Tony Judt, *The New York Review of Books*, May 1, 2003.

Chapter 22. Technology and the Global Ecosystem

1. Flannery, *Eternal Frontier*, 186–93. Ehrlich and Diamond also make this point.
2. An ancient term roughly meaning "mother earth" and recently revived by James Lovelock as an appropriate term applied to the earth's biocosm as a living, integrated entity.
3. Calvin, *A Brain for All Seasons*, 225–9.
4. Burke and Ornstein, *The Axemaker's Gift*, 16–20 and passim.
5. Peter Huber, *Hard Green: Saving the Environment from Environmentalists* (New York: Basic Books, 1999).
6. Ibid., 148.
7. Cass Peterson, Laurie E. Drinkwater, Peggy Wagoner, *Rodale Institute's Farming System Trial* (Kutztown, PA: Rodale, 1999).
8. James Gleick, *Chaos: The Making of a New Science* (New York: Viking, 1987), chap.1.

Chapter 23. A Post-Crash Theocracy for a Stable Global Ecosystem

1. Lionel Tiger and Robin Fox, *The Imperial Animal* (New York: Rinehart and Winston, 1971).
2. An ancient term roughly meaning "mother earth" and recently revived by James Lovelock as an appropriate term applied to the earth's biocosm as a living, integrated entity.

Chapter 24. A More Hopeful Scenario

1. Francis Fukuyama, *Our Posthuman Future: Consequences of the Biotechnology Revolution* (New York: Picador-Farrar, Straus and Giroux, 2002).

2. Isaac Asimov, *Life and Energy* (New York: Doubleday, 1962).
3. Wallace, *Postmodern Management,* chap. 3.
4. Peter Ward and Donald Brownlee, *The Life and Death of Planet Earth* (New York: Henry Holt, 2002). Also, Guillermo Gonzalez and Jay Richards. *The Privileged Planet: How Our Place in the Cosmos is Designed for Discovery* (New York: Regnery, 2004). Chapters 6–10 describe our de-evolution to oblivion.

Appendix 1. Science and Evolution

1. Ehrlich, *Human Natures,* Also, Kingdon, *Self-Made Man.* Both books make this point.
2. Michael J. Behe, *Darwin's Black Box: The Biochemical Challenge to Evolution* (New York: Free Press, 1996), 29–30.
3. Paul Davies and John Gribben, *The Matter Myth* (New York: Simon and Schuster, 1992).
4. Colin Bruce, *Schrödinger's Kittens: The Many Worlds of Quantum* (Washington, DC: Joseph Henry Press, 2004). Colin Bruce offers the latest defense of this theory. Bruce claims that while the many worlds hypothesis originated in the United States its main advocates are now centered in Oxford University in England.

Appendix 2. Autopoiesis

1. Huntington, *Clash of Civilizations.*
2. Ibid., 268.
3. Diamond, *Collapse: How Societies Choose to Fail or Succeed,* chap.2.

SELECTED BIBLIOGRAPHY

Armstrong, Karen. *Islam: A Short History*. New York: Random House, 2000. Reprint, New York: Modern Library Paperback Edition, 2002.

Barzun, Jacques. *From Dawn to Decadence: 500 Years of Western Cultural Life (1500 to the Present*). New York: HarperCollins, 2000.

Behe, Michael J. *Darwin's Black Box: The Biochemical Challenge to Evolution*. New York: Free Press, 1996.

Bruce, Colin. *Schrödinger's Kittens: The Many Worlds of Quantum*. Washington, DC: Joseph Henry Press, 2004.

_______. *Schrödinger's Rabbits: The Many Worlds of Quantum*. Washington, DC: Joseph Henry Press, 2004.

Burke, James, and Robert Ornstein. *The Axemaker's Gift: Technology's Capture and Control of Our Minds and Culture*. New York: Tarcher/Putnam, 1995.

Calvin, William H. *A Brain for All Seasons: Human Evolution and Abrupt Climate Change*. Chicago: University of Chicago Press, 2002.

Cameron, Rondo, and Larry Neal. *A Concise Economic History of the World*. New York: Oxford University Press, 2003.

Cox, W. M., and R. Alm. *Time Well Spent: The Declining Real Cost of Living in America*. Dallas: Federal Reserve Bank of Dallas, 1997.

Anonymous. *Imperial Hubris*. Washington, DC: Brassey's, 2004.

Davies, Paul C., and John Gribbin. *The Matter Myth: Dramatic Discov-*

eries That Challenge Our Understanding of Physical Reality. New York: Simon and Schuster, 1992.

De Tocqueville, Alexis. *Democracy in America.* Washington, DC: Washington Square, 1964.

Diamond, Jared. *The Third Chimpanzee: The Evolution and Future of the Human Animal.* New York: HarperCollins, 1992.

_______. *Guns, Germs, and Steel: The Fates of Human Societies.* New York: Norton 1999.

_______. *Collapse: How Societies Choose to Fail or Succeed.* New York: Viking/Penguin, 2005.

Ehrlich, Paul R. *Human Natures: Genes, Cultures, and the Human Prospect.* Middlesex, England: Penguin, 2002.

Flannery, Tim. *The Eternal Frontier: An Ecological History of North America and Its Peoples.* New York: Atlantic Monthly Press, 2001.

_______. *The Future Eaters.* Australia: Reed New Holland, 1994. Reprint, New York: Grove, 1994.

Freeman, Charles. *Egypt, Greece and Rome: Civilizations of the Ancient Mediterranean.* New York: Oxford University Press, 1996.

Friedman, Thomas L. *The Lexus and the Olive Tree.* New York: Farrar, Straus, and Giroux, 1999.

Fukuyama, Francis. *The End of History and the Last Man.* New York: Oxford University Press, 1986.

_______. *Our Posthuman Future: Consequences of the Biotechnology Revolution.* New York: Farrar, Straus, and Giroux, 2002.

Fuller, Robert C. *Americans and the Unconscious.* New York: Oxford University Press, 1986.

Gleick, James. *Chaos: Making a New Science.* New York: Viking/Penguin, 1987.
Goldsmith, Edward. *The Way: An Ecological World-View.* Athens: Uni-

versity of Georgia Press, 1998.

Gonzalez, Guillermo, and Jay W. Richards. *The Privileged Planet: How Our Place in the Cosmos Is Designed for Discovery*. Washington, DC: Regnery, 2004.

Grant, Michael. *The History of Rome*. New York: Scribner's, 1978.

Gray, D. J. *William Wallace: The King's Enemy*. New York: Barnes and Noble Books, 1991.

Hilton, Isabel. "The Pashtun Code." *The New Yorker* (December 3, 2003): 59-71.

Hobbes, Thomas. *Leviathan,* Richard Tuck, ed. Cambridge: Cambridge University Press, 1998.

Huber, Peter. *Hard Green: Saving the Environment from the Environmentalists*: A *Conservative Manifesto*. New York: Basic Books, 1999.

Huntington, Samuel P. *The Clash of Civilizations and the Remaking of World Order*. New York: Simon and Schuster, 1996.

Judt, Tony. Review of *"The Big Lie,"* by Thierry Meyssan. *New York Review of Books,* 1 May 2003, 7.

Kauffman, Stuart. *Investigations*. New York: Oxford University Press, 2000.

Kingdon, Jonathan, *Self-Made Man: Human Evolution from Eden to Extinction?* New York: Wiley, 1993.

Klein, Richard G., and Edgar, Blake. *The Dawn of Human Culture: A Bold New Theory on What Sparked the "Big Bang" of Human Consciousness*. New York: Wiley, 2002.

Le Goff, Jacques. *Medieval Civilization; 400--1500*. New York: Barnes and Noble Books, 1964.

Lewis, Bernard. *What Went Wrong? The Clash Between Islam and Modernity in the Middle East*. New York: Perennial-HarperCollins, 2002.

Maturana, Humberto, and Francisco Varela. *Autopoiesis and Cognition: The Realization of Learning*. London: Reidl, 1980.

Morton, Frederic. *Thunder at Twilight: Vienna, 1913–1914*. New York: Scribner's, 1989.

Polanyi, Karl. *The Great Transformation*. Boston: Beacon Press, 1944.

Seife, Charles. *Alpha and Omega: The Search for the Beginning and End of the Universe*. New York: Penguin, 2003.

Shire, William L. *Rise and Fall of the Third Reich*. New York: Simon and Schuster, 1959. Reprint, New York: Touchstone, 1981.

Spar, Debra L. *Ruling the Waves: Cycles of Discovery, Chaos, and Wealth from the Compass to the Internet*. New York: Harcourt Brace, 2001.

Tattersall, Ian. *Becoming Human: Evolution and Human Uniqueness*. New York: Harcourt Brace, 1998.

Tiger, Lionel, and Robin Fox. *The Imperial Animal*. New York: Rinehart and Winston, 1971.

Turner, F. J. *The Frontier in American History*. New York: Henry Holt, 1893. Reprint, New York: Henry Holt, 1947.

Wallace, William MacDonald. *How to Save Free Enterprise: From Bureaucrats, Autocrats, and Technocrats*. New York: Dow Jones-Irwin, 1974.

_______. *Postmodern Management: The Emerging Partnership Between Employees and Stockholders*. Westport: Quorum-Greenwood, 1998.

_______. *Supply and Demand Interaction Model*. Seattle: Boeing, 1978.

_______. *TRS-747 Forecast Document*. Seattle: Boeing, 1961.

Walton, Gary M., and Hugh Rockoff. *History of the American Economy, 9th ed*. Australia: South-Western/Thomson Learning, 2002.

Ward, Peter D., and Donald Brownlee. *The Life and Death of Planet Earth: How the New Science of Astrobiology Charts the Ultimate Fate of*

Our World. New York: Henry Holt/Time Books, 2002.

Weber, Max. *General Economic History*. Trans. by Frank H. Knight. New York: Collier, 1961.

_______. *The Protestant Ethic and the Spirit of Capitalism*. Trans. by Talcott Parsons. New York: Scribner's, 1958.

Wheatley, Margaret J. *Leadership and the New Science: Discovering Order in a Chaotic World.* San Francisco: Berrett-Koehler, 1999.

Winik, Jay. *April 1865: The Month That Saved America*. New York: Perennial-HarperCollins, 2002.

Wittfogel, Karl. *Oriental Despotism*. New Haven: Yale University Press, 1963.

INDEX

ABOUT THE AUTHOR

WILLIAM MCDONALD WALLACE is Dean of the Business School at St. Martin's University in Lacey, Washington. He began teaching there in 1992 upon retiring as the Boeing's chief economist for commercial airplanes. He also spent about ten years abroad as a consulting economist. His consulting led him to various nations including the United Kingdom, Brazil, Thailand, Indonesia, Malaysia, Nigeria, and Jordan.

Wallace received his undergraduate degree from the University of the Puget Sound in Tacoma, Washington, and a master's and doctorate degree from the University of Washington in Seattle. He served as a first lieutenant in the Army during the Korean War, most of it in Korea.

Wallace has published two previous books on organization and management in 1974 and 1998 and a number of journal articles. Over the last forty-five years, he has researched the issue of economic growth culture change in relationship to advancing technology.

Wallace lives on a small tree farm in Grays Harbor County in Washington state with his wife, Patricia Ann. He has three children, Patricia Lynn, Scott McDonald, and Coral Ann Fritz, and by the latter two grandchildren, Christopher and Jaclyn.